ESOPHAGEAL DISEASE AND THE ROLE OF THE MICROBIOME

ESOPHAGEAL DISEASE AND THE ROLE OF THE MICROBIOME

Edited by

DAVID A. JOHNSON, MD, MACG, FASGE, MACP

Professor of Medicine/Chief of Gastroenterology,
Eastern Virginia Medical School,
Norfolk, VA, United States

ELSEVIER

ACADEMIC PRESS
An imprint of Elsevier

Academic Press is an imprint of Elsevier
125 London Wall, London EC2Y 5AS, United Kingdom
525 B Street, Suite 1650, San Diego, CA 92101, United States
50 Hampshire Street, 5th Floor, Cambridge, MA 02139, United States
The Boulevard, Langford Lane, Kidlington, Oxford OX5 1GB, United Kingdom

ISBN: 978-0-323-95070-1

For information on all Academic Press publications
visit our website at https://www.elsevier.com/books-and-journals

Publisher: Stacy Masucci
Acquisitions Editor: Osborn M. Patricia
Editorial Project Manager: Billie Jean Fernandez
Production Project Manager: Swapna Srinivasan
Cover Designer: Matthew Limbert

Typeset by STRAIVE, India

Contents

Contributors

Julian A. Abrams, MD, MS
Division of Digestive and Liver Diseases, Department of Medicine, Columbia University Irving Medical Center, New York, NY, United States

Lindsey Cundra, MD
Department of Internal Medicine, Georgetown University, Washington, DC, United States

Steve M. D'Souza, MD
Department of Internal Medicine, Eastern Virginia Medical School, Norfolk, VA, United States

Isaac Davis, MD
Department of Internal Medicine, Eastern Virginia Medical School, Norfolk, VA, United States

Eran Elinav, MD, PhD
Systems Immunology Department, Weizmann Institute of Science, Rehovot, Israel; Microbiome & Cancer Research, DKFZ, Heidelberg, Germany

Alsiddig Elmahdi, MD
Department of Internal Medicine, Eastern Virginia Medical School, Norfolk, VA, United States

C. Prakash Gyawali, MD, MRCP
Division of Gastroenterology, Washington University School of Medicine, Saint Louis, MO, United States

Kevin V. Houston, MD
Department of Internal Medicine, Virginia Commonwealth University, Richmond, VA, United States

David A. Johnson, MD, MACG, FASGE, MACP
Division of Gastroenterology, Department of Internal Medicine, Eastern Virginia Medical School, Norfolk, VA, United States

Philip O. Katz, MD, MACG
Department of Medicine, Division of Gastroenterology and Hepatology, New York Presbyterian—Weill Cornell, New York, NY, United States

David A. Katzka, MD, FACG, AGAF
Columbia University Medical School, Division of Digestive and Liver Disease, New York, NY, United States

Heather Korus, MD
Aurora Health Care, Summit, WI, United States

Michelle Eugene Lee, MD
Department of Medicine, Division of Gastroenterology and Hepatology, New York Presbyterian—Weill Cornell, New York, NY, United States

Alyyah Malick, MD
Division of Digestive and Liver Diseases, Department of Medicine, Columbia University Irving Medical Center, New York, NY, United States

Amit Patel, MD
Division of Gastroenterology, Duke University School of Medicine and the Durham VA Medical Center, Durham, NC, United States

Andrea Proaño-Vasco, PhD
University Medical Center of Freiburg, Clinic for Internal Medicine II, Gastrointestinal Oncology; Spemann Graduate School of Biology and Medicine (SGBM); Faculty of Biology, University of Freiburg, Freiburg, Germany

Michael Quante, MD
University Medical Center of Freiburg, Clinic for Internal Medicine II, Gastrointestinal Oncology; Spemann Graduate School of Biology and Medicine (SGBM), University of Freiburg, Freiburg, Germany

Shabnam Sarker, MD
Department of Medicine, Division of Gastroenterology, Hepatology and Nutrition, Vanderbilt University Medical Center, Nashville, TN, United States

Ali Soroush, MD, MS
Division of Digestive and Liver Diseases, Department of Medicine, Columbia University Irving Medical Center, New York, NY, United States

Rhonda F. Souza, MD, AGAF
Baylor Scott and White Center for Esophageal Diseases, Baylor University Medical Center; Baylor Scott and White Center for Esophageal Research, Baylor Scott and White Research Institute, Dallas, TX, United States

Stuart Jon Spechler, MD, MACG, AGAF
Department of Medicine, Center for Esophageal Diseases, Baylor University Medical Center; Center for Esophageal Research, Baylor Scott & White Research Institute, Dallas, TX, United States

Igor Spivak, MD
Systems Immunology Department, Weizmann Institute of Science, Rehovot, Israel; Medical Clinic III, University Hospital Aachen, Aachen, Germany

Michael F. Vaezi, MD, PhD, FACP, AGAF
Department of Medicine, Division of Gastroenterology, Hepatology and Nutrition, Vanderbilt University Medical Center, Nashville, TN, United States

Nimish Vakil, MD, FACG, AGAF
University of Wisconsin School of Medicine and Public Health, Madison, WI, United States

Manasa Vallabhaneni, MD
Department of Internal Medicine, Eastern Virginia Medical School, Norfolk, VA, United States

Ana Vilela, MD
Department of Internal Medicine, Eastern Virginia Medical School, Norfolk, VA, United States

Byung Soo Yoo, MD
Department of Internal Medicine, Eastern Virginia Medical School, Norfolk, VA; Division of Gastroenterology, Department of Internal Medicine, Carolinas Medical Center, Charlotte, NC, United States

Introduction

David A. Johnson, MD, MACG, FASGE, MACP
Professor of Medicine/Chief of Gastroenterology, Eastern Virginia Medical School, Norfolk, VA, United States

The human gastrointestinal (GI) tract contains a complex array of bacteria, fungi (mostly yeasts), viruses, and archaea. This diverse collection of microorganisms, typically, is referred to as the "gut microbiota" and the genes involved as the "gut microbiome." The components of each in fact contain trillions of cells and genetic material that far exceeds the host registry. These populations interact with the human genotype and environment to maintain body homeostasis or perhaps, as suggested, to initiate pathogenic diseases, when the balance is altered. This "dysbiosis" may involve loss of beneficial microbes, decreased diversity, or pathobiont expansion. Consequent effects can include, but are not limited to, adverse alterations of immune function, epithelial barrier function, gene activation, inflammatory and neoplastic pathways, motility, and nutrient absorption. This ever-increasing recognition has led to the evaluation of a possible microbial role in the pathogenesis of and ever-expanding list of both GI and non-GI diseases.

In the GI tract, a pivotal microbial role perhaps is exemplified by the role of *Helicobacter pylori* in peptic ulcer disease and gastric cancer. More recent evidence has suggested a prominent role in many common other inflammatory and neoplastic diseases, in particular, including the liver, pancreatic-biliary, small intestine, and colon. Notably missing from the compendiums focusing on these disease states are the ever-increasing data on the reported associations of dysbiosis-related associations with very prevalent esophageal diseases, ranging from gastroesophageal reflux disease (GERD) and consequences to other inflammatory and motility disorders.

Lack of information is clearly an impediment to better understanding and thereby critical for expanding understanding for best approaches to disease management. Translational research begins with scientific understanding/hypothesis generation. The goal of this treatise therefore is to educate providers on the rapidly emerging evidence on the *potential* biomic influence on esophageal diseases. Recognizably, the most evidence-based confirmation of causality would be a defined causative pathway and clearly defined response to a directed mitigation strategy. Unfortunately, at present, we cannot offer this level of evidence. Dr. Don Castell, heretofore recognized as the "godfather of esophagology," an esteemed mentor as well as valued friend to so many of the present authors, would say "That a *good* question … hopefully, will result in a *good* answer … however, a *great* question … will not only provide a ***great*** **answer** … but more importantly … inspire 10 more ***great questions*** of what is needed next to provide more 'good answers' …"

The aim of this book therefore is to provide the most "cutting-edge" science on associations, technologic assessments, and perhaps, where appropriate, data addressing possible mitigation strategies. I am indebted, with the utmost thanks, to my internationally acclaimed experts in this area, to present not the answers, but perhaps the challenge … based on the emerging translational evidence … and … to ask thereby the readers to inspire translational efforts by asking more ***great questions*** … en route to ***great answers***!

CHAPTER 1

Is acid the cornerstone factor in gastroesophageal reflux disease and its related complications?

Stuart Jon Spechler[a,b]
[a]Department of Medicine, Center for Esophageal Diseases, Baylor University Medical Center, Dallas, TX, United States
[b]Center for Esophageal Research, Baylor Scott & White Research Institute, Dallas, TX, United States

Introduction—A brief history of gastric acid

The role of gastric acid in gastrointestinal disorders has been a subject of great interest and intense controversy for hundreds of years. In the early 16th century, the Swiss alchemist/physician Paracelsus proposed that the stomach of certain animals contained acid that participated in the digestion of food [1]. However, Paracelsus did not think that the human stomach made acid, and suggested that humans would have acid in their stomachs only if they drank acidic spa waters. Later in that century, the Flemish chemist/physician Jean Baptiste van Helmont also recognized that animals had acid in the stomach, and also suggested that acid played a vital role in digestion. Since he observed that "neither doth vinegar nor the broth of citron leaven nor ferment the meal" (i.e., he observed that acid alone in vitro did not digest food), he proposed that another factor, the "ferment," was also required (along with acid) for the process of digestion. In the 17th century, the German-born, Dutch physician Franciscus Sylvius (founder of the Iatrochemical School of Medicine, which held that all life and disease processes were the results of chemical reactions) proposed that swallowed saliva initiated digestion in the stomach, and that the pancreas, not the stomach, supplied acid that mixed with bile to digest food [1].

The iatrochemists' contention that acid played a key role in digestion was opposed by iatrophysicists in 17th and 18th centuries who believed that vital processes and diseases could be explained by physicomechanical rather than chemical principles. The Italian physiologist/mathematician Giovanni Borelli considered that digestion is a mechanical process initiated by the physical force of gastric contractions, and the Dutch physician Herman Boerhaave

Esophageal Disease and the Role of the Microbiome
https://doi.org/10.1016/B978-0-323-95070-1.00012-6

rejected the notion that acid played a key role in digestion. The 18th-century Swiss physiologist Albrecht von Haller contended that the stomach did not produce acid, but rather that any acid found in the stomach was merely the product of retained food fermenting in the stomach. For researchers in the 18th century, studies on the stomach's role in digestion typically involved gastric juice and digested food products recovered from animals or vomited up voluntarily by human researchers. These studies were not definitive and often reached contradictory conclusions. Consequently, the issue of whether acid in the stomach was a gastric secretory product or a fermentation product of retained ingested material remained hotly debated until well into the 19th century.

In 1823, the English chemist/physician William Prout definitively identified hydrochloric acid (also called muriatic acid) as a major constituent of gastric juice [2]. This observation was confirmed in seminal human experiments conducted by William Beaumont, a US Army surgeon [3,4]. This is an interesting story. In June 1822, while stationed on the Island of Michilimackinac (now called Mackinac Island) in the territory that is now Michigan, Beaumont was called to treat a patient named Alexis St. Martin, a 19-year-old French Canadian fur trapper who had taken an extensive shotgun wound to the abdomen. Beaumont's description of his first encounter with St. Martin is horrifying: "*I was called to him immediately after the accident. Found a portion of the Lungs as large as a turkey's egg protruding through the external wound, lacerated and burnt, and below this another protrusion resembling a portion of the Stomach, what at first view I could not believe possible to be that organ in that situation with the subject surviving, but on closer examination I found it to be actually the Stomach, with a puncture in the protruding portion large enough to receive my forefinger, and through which a portion of his food that he had taken for breakfast had come out and lodged among his apparel. In this dilemma I considered my attempt to save his life entirely useless.*" Miraculously, St. Martin did survive until the age of 83, but with a gastric fistula that allowed Beaumont (unhampered by institutional review boards) easy access into the gastric lumen so that he could sample its contents and observe firsthand the ability of gastric juice to digest food both inside and outside the stomach. In 1833, Beaumont published a book entitled "Experiments and Observations on the Gastric Juice and the Physiology of Digestion" in which he described his 238 experiments on St. Martin. In an address to the St. Louis Medical Society in 1902, William Osler gave Beaumont credit for confirming Prout's observation on the importance of hydrochloric acid in gastric juice and for his "refutation of many erroneous opinions relating to gastric digestion" [4].

In 1834, the German physiologist Johann Eberle noted that weak hydrochloric acid and neutral gastric mucosal extracts applied individually to organic substances had little effect but, when mixed together, they rapidly dissolved organic material. Based on this observation he concluded, as had van Helmont more than two centuries earlier, that there was a "ferment" that contributed to digestion. In 1835, Theodor Schwann, another German physiologist, isolated from gastric mucosa the enzyme responsible for this ferment, which he named pepsin (from the Greek word "pepsis" meaning digestion). In 1882, Quincke coined the term "peptic ulcer," for ulcers that he believed were caused by the proteolytic effects of pepsin and the corrosive effects of gastric acid [5,6]. In 1910, surgeon Karl Schwarz (born in what today is Croatia) published a series of 14 patients with ulcers in the upper GI tract that he blamed on exposure to gastric acid, proclaiming his dictum "no acid, no ulcer" [7].

In 1935, Asher Winkelstein, a gastroenterologist practicing at Mt. Sinai Hospital in New York, extended these concepts of acid-peptic damage to the esophagus in a report describing a series of patients who had heartburn associated with endoscopic and histologic signs of inflammation in the distal esophagus [8]. He proposed that those patients had "peptic esophagitis ... resulting from the irritant action on the mucosa of free hydrochloric acid and pepsin." Winkelstein's concept that GERD develops as an acid-peptic "burn" of the esophagus went largely unchallenged until 2009 when Rhonda Souza and I proposed that acid incites esophageal injury through a cytokine-mediated mechanism [9]. According to this hypothesis, refluxed gastric acid does not kill esophageal squamous cells directly, but rather stimulates them to produce pro-inflammatory cytokines, which attract the inflammatory cells that ultimately damage the esophagus. Dr. Souza discusses this hypothesis in detail in another chapter of this book.

Evidence that acid causes esophageal injury

Acid is not the only potentially damaging agent for the esophagus in refluxed gastric juice. Refluxed bile acids and pancreatic enzymes also might contribute to the development of reflux esophagitis (see below). Before the advent of proton pump inhibitors (PPIs), which have been available for clinical use in the United States since 1989, experts often debated the relative importance of acid, bile, and pancreatic enzymes in the pathogenesis of reflux esophagitis. Those debates seemed unnecessary and virtually ceased when it was appreciated that reflux esophagitis heals in the large majority of patients treated with the potent gastric acid suppression that can be achieved by PPI therapy. Indeed, such debates would be superfluous if acid suppression were the only mechanism whereby PPIs might heal reflux esophagitis, which remains a common misconception. Far less appreciated than the antisecretory effects of the PPIs are their antiinflammatory actions, which are independent of their effects on gastric acid secretion and which also might contribute importantly to the healing of GERD. For example, PPIs have antioxidant properties, they have inhibitory effects on neutrophil function, they can decrease adhesion molecule production by endothelial cells and neutrophils, and they can decrease the production of pro-inflammatory cytokines by endothelial and epithelial cells [10,11]. Thus, the response of a disease to PPI therapy does not establish acid as the only potential pathogenetic factor.

Direct experimental evidence has established that acid alone can damage the esophagus, but only under certain conditions. Concentrated acid is extremely corrosive, and the ingestion of concentrated acid (e.g., the sulfuric acid found in car batteries, which has a pH level of 0.8) can cause severe caustic injury to the esophagus and stomach [12]. However, the production of gastric juice with pH levels <1 is beyond the secretory capacity of the normal human stomach [13,14]. In a study of 154 GERD patients who underwent 24-hour esophageal pH monitoring, Bremner et al. found that the large majority of reflux episodes had pH levels >2, but patients whose esophageal exposure to pH levels <2 exceeded 1.32% (the 95th percentile in this study) had the highest prevalence of reflux esophagitis and Barrett's esophagus [15].

Goldberg et al. reported that perfusion of the cat esophagus for 1 hour with HCl at pH levels ≤ 1.3 caused esophagitis, but HCl alone at pH levels >1.3 did not cause esophagitis [16]. However, esophagitis was produced at higher pH levels (1.6 and 2.0) when pepsin was added to the perfusate, but not at pH levels >2.3. Johnson and Harmon noted similar results in studies in which perfusion of the rabbit esophagus with HCl at physiologic pH values did not cause esophagitis, whereas perfusions with pepsin in an acid environment (pepsin has its greatest proteolytic activity at pH levels ≤ 2 [17], and its proteolytic activity virtually ceases at pH levels >4) and trypsin in an alkaline environment (which might be found in the refluxate of patients taking PPIs) caused severe esophagitis [14]. Those investigators concluded that the importance of acid in the pathogenesis of reflux esophagitis had been overemphasized, while the role of pepsin had been underappreciated and minimized. That observation may still hold true today.

Acid and heartburn

Heartburn, the cardinal symptom of GERD, typically is described as a burning sensation in the retrosternal area that rolls up the chest. It is often precipitated by foods that induce reflux like chocolate and fats, or by foods that might irritate a GERD-damaged esophagus like citrus fruits and tomato products. Furthermore, the heartburn of GERD typically is eased temporarily by antacids and eliminated with antisecretory medications, features often taken to indicate that refluxed acid is the primary agent responsible for causing the sensation of heartburn. Traditionally, it has been assumed that heartburn results when refluxed gastric acid contacts a reflux-damaged esophageal mucosa [18]. Since reflux esophagitis impairs esophageal mucosal integrity, heartburn has been explained by refluxed protons penetrating the leaky mucosa and stimulating nociceptors (nerves that transmit the sensation of heartburn) by activating their acid-sensing ion channel (ASIC) receptors or their transient receptor potential vanilloid-1 (TRPV1) receptors [18,19].

As conceptually appealing as this traditional explanation for the pathogenesis of heartburn might be, it does not readily account for several observations. For example, the correlation between episodes of acid reflux and episodes of heartburn in esophageal pH monitoring studies of GERD patients often is very poor, with acid reflux frequently occurring in the absence of heartburn and vice versa [20]. In a study of healthy volunteers undergoing unsedated endoscopy, the injection of acid into the esophageal mucosa often elicited no symptoms, whereas injection of the TRPV1 agonist capsaicin caused severe heartburn [21]. Perfusion of the esophagus with menthol, an activator of transient receptor potential melastatin 8 (TRPM8) receptors, has been shown to elicit an innocuous cold sensation in healthy volunteers, but a painful heartburn sensation in GERD patients [22]. These studies clearly demonstrate that acid is not the only substance which can elicit the sensation of heartburn from the esophagus, esophageal acid exposure does not always result in heartburn, and heartburn can occur in the absence of acid exposure. They also suggest that GERD patients might have alterations in esophageal afferent nerve signaling such that stimuli other than acid induce the sensation of heartburn.

By convention, esophageal pH monitoring identifies an episode of acid reflux only when esophageal pH falls below four. In contrast, esophageal impedance monitoring identifies all reflux episodes whether they are acidic ($pH < 4$), weakly acidic ($pH = 4–7$), or alkaline ($pH > 7$). This is because impedance monitoring measures the electrical resistance between electrodes on a catheter positioned in the esophagus. Any liquid that refluxes across the electrodes causes a sudden drop in electrical resistance irrespective of the liquid's pH level, whereas the reflux of gas (as in a belch) results in a precipitous rise in electrical resistance. Studies using esophageal impedance-pH monitoring have demonstrated that the reflux of nonacidic material or gas both can evoke the sensation of heartburn [23,24].

It is not clear why the reflux of nonacidic material or gas causes heartburn. Perhaps, weakly acidic material activates the same nociceptors that strongly acidic material is assumed to activate. Alternatively, there might be irritants other than acid in gastric juice such as bile salts that can activate nociceptors. Mechanical distention of the esophagus with a balloon has been shown to elicit heartburn [25], and thus, it is conceivable that large volume reflux of fluid or gas might cause sufficient distention to activate heartburn-conveying nociceptors. In support of the latter concept, studies have shown that refluxed material which reaches the proximal esophagus is more likely to be associated with heartburn than reflux confined to the distal esophagus [24]. Since the level of extension of refluxed material into the esophagus appears to be directly related to refluxate volume, material that reaches the proximal esophagus might have sufficient volume to cause the distention that triggers esophageal nociceptors [26].

Studies in which high-frequency intraluminal ultrasonography was used to image the esophageal wall of patients with heartburn found that their heartburn episodes were associated with sustained contractions of esophageal longitudinal muscle [27]. Such sustained esophageal contractions could result in localized ischemia, which might trigger nociceptors to convey the sensation of heartburn [28,29]. As discussed earlier, reflux esophagitis appears to develop through a cytokine-mediated process in which refluxed gastric material induces esophageal epithelial cells to secrete pro-inflammatory cytokines [9]. The resulting inflammation might be an acid-independent trigger for heartburn. Alternatively, the pro-inflammatory cytokines themselves might trigger heartburn. Finally, acid reflux is associated with a change in the esophageal microbiome from one predominated by gram-positive organisms to one with an increased population of gram-negative bacteria that have a lipopolysaccharide (LPS) shell, which might protect them from destruction by refluxed acid [30]. LPS can activate Toll-like receptor pathways in esophageal epithelial cells that promote the production of pro-inflammatory cytokines [31,32], and LPS sensitizes TRPV1 receptors, which might convey the sensation of heartburn [33]. Thus, refluxed acid might play an indirect role in the pathogenesis of GERD by inducing an esophageal dysbiosis that can trigger inflammation and the sensation of heartburn and, conceivably might predispose the esophageal mucosa to further damage from acid reflux. The topic of esophageal dysbiosis is covered in detail elsewhere in this book. In summary, despite many decades of intense research, the mechanisms underlying the heartburn of GERD remain unclear, and acid is not the only factor.

Bile acids, esophageal injury, and heartburn

Bile acids are steroid-like molecules produced in the liver from cholesterol [34]. The predominant bile acids in human bile are cholic acid and chenodeoxycholic acid (the primary bile acids synthesized by the liver), and deoxycholic acid (a secondary bile acid produced within the intestinal lumen by bacterial dehydroxylation of cholic acid). Bacteria in the gut also convert chenodeoxycholic acid into lithocholic acid, but this secondary bile acid is present only in very low concentrations in human bile. Free (unconjugated) bile acids, which have pK_a values of 5–7, are hydrophobic and poorly soluble at gastric pH levels. Before the liver secretes bile acids into bile, it conjugates them with glycine or taurine, which lowers their pK_a (taurine conjugates pK_a value of 2, whereas glycine conjugates pK_a value of 4) and renders them more soluble in gastric juice. Approximately 98% of bile acids in bile are in this conjugated form [35].

A recent study has shown that deoxycholic acid can directly activate and sensitize nociceptive nerves in the guinea pig esophagus, suggesting that bile acids might contribute to the production of heartburn and chest pain in GERD patients [36]. Other animal studies have established that the reflux of bile alone (i.e., without gastric acid reflux) can cause severe esophagitis [37–39], but human studies on the role of bile acids in the pathogenesis of reflux esophagitis are less clear cut. In human esophageal cells, exposure to bile acids can increase the expression of reactive oxygen species and proinflammatory cytokines [11,40–42]. Compared to control subjects, patients with GERD have higher concentrations of bile in refluxed gastric juice, and gastric juice bile acid concentrations are especially high in those with severe esophagitis (LA grades C and D) and Barrett's esophagus [43–45]. Studies in which bile acids alone or together with HCl solutions were perfused into the esophagus of awake patients have shown that bile acids can induce heartburn and chest discomfort, especially when they are combined with HCl [46,47]. Bile acids can also induce DNA damage in esophageal cells, and can cause esophageal squamous cells to express intestinal-like genes (e.g., CDX2,

BMP4, MUC2) similar to those found in Barrett's metaplasia [48–52]. These latter features, which suggest a role for refluxed bile acids in the pathogenesis of Barrett's esophagus and its progression to esophageal adenocarcinoma, are beyond the scope of this chapter.

In the absence of surgery that removes or diverts the stomach, the esophagus is exposed to bile acids only in refluxed gastric juice, which is usually acidic. This makes it difficult to determine the precise role of bile acids alone (without acid) in the production of GERD signs and symptoms. Nevertheless, the data discussed earlier suggest that bile acids contribute, along with acid, to the pathogenesis of reflux esophagitis and the induction of heartburn in GERD patients.

Conclusions

This chapter was intended to answer the question "Is acid the cornerstone factor in GERD and its related complications?" Although I believe that the answer is yes, the precise role of acid in the pathogenesis of GERD and its complications remains unclear. Acid alone does not appear to be capable of damaging the esophagus unless it has a pH of 1.3 or less, a level uncommonly observed during reflux episodes in GERD patients. Acid-peptic damage can occur at higher pH levels (perhaps up to ~4), but recent studies suggest that acid-peptic attack usually does not kill esophageal cells directly, but rather it induces them to secrete pro-inflammatory cytokines that attract inflammatory cells, which ultimately kill the esophageal cells. The role of refluxed acid and bile in causing the sensation of heartburn also remains unclear. While it is clear that acid perfusion of the esophagus can elicit heartburn, it is also clear that acid is not the only substance that can do so, and that heartburn can occur in the absence of acid exposure. PPIs have been very successful in treating GERD for most patients, but PPIs have actions independent of gastric acid suppression that might contribute to their beneficial effects. Finally, pathologic acid reflux appears to induce an esophageal dysbiosis that can contribute to esophageal inflammation and heartburn, and that might predispose the esophageal mucosa to further damage from acid reflux. While our understanding of the role acid plays in the pathogenesis of GERD and its complications has evolved considerably over the years, much remains to be learned. Acid plays a key role in GERD, but that role is not as clear-cut as that envisioned by Asher Winkelstein in 1935, and the success of PPIs in treating GERD did not close the chapter on the role of acid in the esophagus.

Future directions

A clearer understanding of the precise roles of refluxed acid and bile, the esophageal cytokines they induce, and acid-induced effects on esophageal dysbiosis in the pathogenesis of reflux esophagitis and the production of heartburn might lead to better treatments for GERD patients. Future studies that might lead to clearer understanding are as follows:

- Studies on the role of individual cytokines in producing reflux esophagitis and heartburn.
- Studies involving cytokine inhibitors in the healing of reflux esophagitis and heartburn.
- Studies characterizing the effects of acid and bile salts individually in altering the esophageal microbiome.
- Studies on targeted therapies toward mitigating effects on the esophageal microbiome and heartburn in GERD patients.
- Studies on how microbiome alterations influence the sensation of heartburn induced by esophageal perfusion of acid and bile salts.

Conflict of interest disclosures

Dr. Spechler is a consultant for Phathom Pharmaceuticals, Ironwood Pharmaceuticals, Takeda Pharmaceuticals, Castle Biosciences, Interpace Diagnostics, and ISOThrive.

References

[1] J.H. Baron, The discovery of gastric acid, Gastroenterology 76 (1979) 1056–1064.
[2] W. Prout, On the nature of the acid and saline matters usually existing in the stomach of animals, Philos. Trans. R. Soc. 1 (1824) 45–49 (Read December 11, 1823).
[3] J.S. Myers, Life and Letters of Dr. William Beaumont, C. V. Mosby, St. Louis, 1912.
[4] W. Osler, William Beaumont. A pioneer American physiologist, JAMA 39 (1902) 1223–1231.
[5] A.D. Bevan, Peptic ulcer: etiology, history and surgical treatment, JAMA 94 (1930) 2043–2046.
[6] S.X.M. Dong, C.C.Y. Chang, K.J. Rowe, A collection of the etiological theories, characteristics, and observations/phenomena of peptic ulcers in existing data, Data Brief 19 (2018) 1058–1067.
[7] J. Gustafson, D. Welling, "No acid, no ulcer"—100 years later: a review of the history of peptic ulcer disease, J. Am. Coll. Surg. 210 (2010) 110–116.
[8] A. Winkelstein, Peptic esophagitis: a new clinical entity, JAMA 104 (11) (1935) 906–909.
[9] R.F. Souza, X. Huo, V. Mittal, et al., Gastroesophageal reflux may cause esophagitis through a cytokine-mediated mechanism, not by caustic (acid) injury, Gastroenterology 137 (5) (2009) 1776–1784.

[10] R.R. Kedika, R.F. Souza, S.J. Spechler, Potential anti-inflammatory effects of the proton pump inhibitors: a review and discussion of the clinical implications, Dig. Dis. Sci. 54 (2009) 2312–2317.
[11] X. Huo, X. Zhang, C. Yu, et al., In oesophageal squamous cells exposed to acidic bile salt medium, omeprazole inhibits IL-8 expression through effects on nuclear factor-κB and activator protein-1, Gut 63 (2014) 1042–1052.
[12] S.A. Zargar, R. Kochhar, B. Nagi, S. Mehta, S.K. Mehta, Ingestion of corrosive acids. Spectrum of injury to upper gastrointestinal tract and natural history, Gastroenterology 97 (1989) 702–707.
[13] E.W. Moore, R.W. Scarlata, The determination of gastric acidity by the glass electrode, Gastroenterology 49 (1965) 178–188.
[14] L.F. Johnson, J.W. Harmon, Experimental esophagitis in a rabbit model. Clinical relevance, J. Clin. Gastroenterol. 8 (Suppl. 1) (1986) 26–44.
[15] R.M. Bremner, P.F. Crookes, T.R. DeMeester, J.H. Peters, H.J. Stein, Concentration of refluxed acid and esophageal mucosal injury, Am. J. Surg. 164 (1992) 522–526.
[16] H.I. Goldberg, W.J. Dodds, S. Gee, C. Montgomery, F.F. Zboralske, Role of acid and pepsin in acute experimental esophagitis, Gastroenterology 56 (1969) 223–230.
[17] Q. Luo, D. Chen, R.M. Boom, A.E.M. Janssen, Revisiting the enzymatic kinetics of pepsin using isothermal titration calorimetry, Food Chem. 268 (2018) 94–100.
[18] W.J. Barlow, R.C. Orlando, The pathogenesis of heartburn in nonerosive reflux disease: a unifying hypothesis, Gastroenterology 128 (2005) 771–778.
[19] P. Holzer, Acid-sensitive ion channels and receptors, Handb. Exp. Pharmacol. 194 (2009) 283–332.
[20] F. Baldi, F. Ferrarini, A. Longanesi, M. Ragazzini, L. Barbara, Acid gastroesophageal reflux and symptom occurrence. Analysis of some factors influencing their association, Dig. Dis. Sci. 34 (1989) 1890–1893.
[21] R.H. Lee, H. Korsapati, V. Bhalla, N. Varki, R.K. Mittal, Esophageal submucosal injection of capsaicin but not acid induces symptoms in normal subjects, J. Neurogastroenterol. Motil. 22 (2016) 436–443.
[22] P. Banovcin, M. Duricek, T. Zatko, P. Liptak, R. Hyrdel, M. Kollarik, The infusion of menthol into the esophagus evokes cold sensations in healthy subjects but induces heartburn in patients with gastroesophageal reflux disease (GERD), Dis. Esophagus 32 (11) (2019). doz038.
[23] A. Agrawal, J. Roberts, N. Sharma, R. Tutuian, M. Vela, D.O. Castell, Symptoms with acid and nonacid reflux may be produced by different mechanisms, Dis. Esophagus 22 (2009) 467–470.
[24] A.J. Bredenoord, B.L. Weusten, W.L. Curvers, R. Timmer, A.J. Smout, Determinants of perception of heartburn and regurgitation, Gut 55 (2006) 313–318.
[25] K.C. Trimble, A. Pryde, R.C. Heading, Lowered oesophageal sensory thresholds in patients with symptomatic but not excess gastro-oesophageal reflux: evidence for a spectrum of visceral sensitivity in GORD, Gut 37 (1995) 7–12.
[26] P.W. Weijenborg, A.J. Bredenoord, How reflux causes symptoms: reflux perception in gastroesophageal reflux disease, Best Pract. Res. Clin. Gastroenterol. 27 (2013) 353–364.
[27] N. Pehlivanov, J. Liu, R.K. Mittal, Sustained esophageal contraction: a motor correlate of heartburn symptom, Am. J. Physiol. Gastrointest. Liver Physiol. 281 (2001) G743–G751.
[28] R.K. Mittal, V. Bhargava, H. Lal, Y. Jiang, Effect of esophageal contraction on esophageal wall blood perfusion, Am. J. Physiol. Gastrointest. Liver Physiol. 301 (2011) G1093–G1098.
[29] Y. Jiang, V. Bhargava, Y.S. Kim, R.K. Mittal, Esophageal wall blood perfusion during contraction and transient lower esophageal sphincter relaxation in humans, Am. J. Physiol. Gastrointest. Liver Physiol. 303 (2012) G529–G535.
[30] L. Yang, X. Lu, C.W. Nossa, F. Francois, R.M. Peek, Z. Pei, Inflammation and intestinal metaplasia of the distal esophagus are associated with alterations in the microbiome, Gastroenterology 137 (2009) 588–597.

[31] S.M. D'Souza, K. Houston, L. Keenan, B.S. Yoo, P.J. Parekh, D.A. Johnson, Role of microbial dysbiosis in the pathogenesis of esophageal mucosal disease: a paradigm shift from acid to bacteria? World J. Gastroenterol. 27 (18) (2021) 2054–2072.
[32] S.M. D'Souza, L.B. Cundra, B.S. Yoo, P.J. Parekh, D.A. Johnson, Microbiome and gastroesophageal disease: pathogenesis and implications for therapy, Ann. Clin. Gastroenterol. Hepatol. 4 (2020) 020–033.
[33] A. Cauwels, J. Bultinck, R. De Zwaef, B. Vandendriessche, S. Magez, P. Brouckaert, Nitric oxide production by endotoxin preparations in TLR4-deficient mice, Nitric Oxide 36 (2014) 36–43.
[34] A.F. Hofmann, L.R. Hagey, Bile acids: chemistry, pathochemistry, biology, pathobiology, and therapeutics, Cell. Mol. Life Sci. 65 (2008) 2461–2483.
[35] K.R. McQuaid, L. Laine, M.B. Fennerty, R. Souza, S.J. Spechler, Systematic review: the role of bile acids in the pathogenesis of gastro-oesophageal reflux disease and related neoplasia, Aliment. Pharmacol. Ther. 34 (2011) 146–165.
[36] X. Yu, Y. Hu, M. Yu, B.J. Undem, S. Yu, Deoxycholic acid activates and sensitizes vagal nociceptive afferent C-fibers in guinea pig esophagus, Am. J. Physiol. Gastrointest. Liver Physiol. 321 (2021) G149–G156.
[37] A.P. Ireland, J.H. Peters, T.C. Smyrk, et al., Gastric juice protects against the development of esophageal adenocarcinoma in the rat, Ann. Surg. 224 (1996) 358–370.
[38] K. Nishijima, K. Miwa, T. Miyashita, et al., Impact of the biliary diversion procedure on carcinogenesis in Barrett's esophagus surgically induced by duodenoesophageal reflux in rats, Ann. Surg. 240 (2004) 57–67.
[39] A.T. Agoston, T.H. Pham, R.D. Odze, et al., Columnar-lined esophagus develops via wound repair in a surgical model of reflux esophagitis, Cell. Mol. Gastroenterol. Hepatol. 6 (2018) 389–404.
[40] A.J. Jolly, C.P. Wild, L.J. Hardie, Sodium deoxycholate causes nitric oxide mediated DNA damage in oesophageal cells, Free Radic. Res. 43 (2009) 234–240.
[41] L.A. Feagins, H.Y. Zhang, X. Zhang, et al., Mechanisms of oxidant production in esophageal squamous cell and Barrett's cell lines, Am. J. Physiol. 294 (2008) G411–G417.
[42] N. Yoshida, E. Imamoto, K. Uchiyama, et al., Molecular mechanisms involved in interleukin-8 production by normal human oesophageal epithelial cells, Aliment. Pharmacol. Ther. 24 (2006) 219–226.
[43] D. Nehra, P. Howell, C.P. Williams, et al., Toxic bile acids in gastro-oesophageal reflux disease: influence of gastric acidity, Gut 44 (1999) 598–602.
[44] D. Nehra, P. Howell, J.K. Pye, et al., Assessment of combined bile acid and pH profiles using an automated sampling device in gastro-oesophageal reflux disease, Br. J. Surg. 85 (1998) 134–137.
[45] W.K. Kauer, J.H. Peters, T.R. DeMeester, et al., Composition and concentration of bile acid reflux into the esophagus of patients with gastroesophageal reflux disease, Surgery 122 (1997) 874–881.
[46] A. Siddiqui, S. Rodriguez-Stanley, S. Zubaidi, et al., Esophageal visceral sensitivity to bile salts in patients with functional heartburn and in healthy control subjects, Dig. Dis. Sci. 50 (2005) 81–85.
[47] G.S. Bachir, J.L. Collis, Effect of perfusion of bile salts solutions into the oesophagus of hiatal hernia patients and controls, Thorax 31 (1976) 271–277.
[48] G. Zhou, Y.G. Sun, H.B. Wang, et al., Acid and bile salt up-regulate BMP4 expression in human esophageal epithelium cells, Scand. J. Gastroenterol. 44 (2009) 926–932.
[49] A.H.C. van Roon, G.C. Mayne, B.P. Wijnhoven, et al., Impact of gastro-esophageal reflux on mucin mRNA expression in the esophageal mucosa, J. Gastrointest. Surg. 12 (2008) 1331–1340.
[50] Y. Hu, C. Jones, O. Gellersen, et al., Pathogenesis of Barrett esophagus: deoxycholic acid up-regulates goblet-specific gene MUC2 in concert with CDX2 in human esophageal cells, Arch. Surg. 142 (2007) 540–544.

[51] S. Peng, X. Huo, D. Rezaei, et al., In Barrett's esophagus patients and Barrett's cell lines, ursodeoxycholic acid increases antioxidant expression and prevents DNA damage by bile acids, Am. J. Physiol. Gastrointest. Liver Physiol. 307 (2014) G129–G139.
[52] X. Huo, K.B. Dunbar, X. Zhang, Q. Zhang, S.J. Spechler, R.F. Souza, In Barrett's epithelial cells, weakly acidic bile salt solutions cause oxidative DNA damage with response and repair mediated by p38, Am. J. Physiol. Gastrointest. Liver Physiol. 318 (3) (2020) G464–G478.

CHAPTER 2

Cytokines and immune modulators—Effects on esophageal inflammation

Rhonda F. Souza[a,b]
[a]Baylor Scott and White Center for Esophageal Diseases, Baylor University Medical Center, Dallas, TX, United States
[b]Baylor Scott and White Center for Esophageal Research, Baylor Scott and White Research Institute, Dallas, TX, United States

Introduction

Traditionally, the pathogenesis of reflux esophagitis (RE) has been viewed as a caustic, chemical injury (i.e., an acid burn). This traditional concept began in 1935, when a gastroenterologist named Asher Winkelstein described a series of patients with heartburn and distal esophageal inflammation, and he proposed that these patients had "peptic esophagitis...resulting from the irritant action on the mucosa of reflux of gastric hydrochloric acid and pepsin" [1]. According to this hypothesis, exposure of squamous epithelium to acid and pepsin, the main noxious constituents of gastric refluxate, are thought to damage the junctions between squamous cells at the luminal surface, making the epithelium permeable and allowing acid to enter and attack the epithelial cells [2,3]. Death of surface cells triggers the infiltration of granulocytes, including neutrophils and eosinophils, and induces a proliferative response resulting in basal cell and papillary hyperplasia to repair the injured epithelium [4–8]. More recently, Winkelstein's fundamental concept that RE develops as an acid-peptic "burn" of the esophageal mucosa has been challenged by studies in rats and humans, demonstrating an immune-mediated inflammatory response [9,10]. This chapter will review the esophageal mucosal defense systems, the role of cytokines as initiators of acute reflux esophagus, and the contribution of the innate and adaptive immune systems to reflux-induced esophageal inflammation.

Esophageal mucosal defense systems

The esophagus is a hollow, fibromuscular tube with a network of organized strata, providing safe passage of nutrients into the stomach and protecting

Esophageal Disease and the Role of the Microbiome
https://doi.org/10.1016/B978-0-323-95070-1.00009-6

against the harsh gastric contents during gastroesophageal reflux. Despite these recurring insults, the normal esophagus remains largely uninjured due to pre-epithelial, epithelial, and post-epithelial defense systems. Esophageal pre-epithelial defense is meagre in comparison to the stomach and duodenum because the esophagus lacks a thick mucus coat on its luminal surface, the surface epithelial cells do not secrete bicarbonate ions to buffer hydrogen ions in the gastric refluxate, and the tightness of the epithelial barrier blocks diffusion of bicarbonate from the blood to the luminal surface [11]. Post-epithelial defense involves components of the lamina propria, including endothelial cells, fibroblasts, resident T cells, and myofibroblasts, a subpopulation of stromal cells subjacent to the epithelial basement membrane. Although post-epithelial defense has been mostly attributed to increases in blood supply in response to luminal acid exposure, recent data suggest that myoepithelial cells, and perhaps resident T cells, contribute to this defense system [12–17]. In fact, esophageal myofibroblasts have been reported to secrete factors including Gremlin and bone morphogenetic proteins, which regulate growth of the squamous epithelium in a paracrine fashion [18]. However, the most important contributor to protection of the esophagus is the epithelial defense system.

The nonkeratinized, stratified squamous epithelium of the human esophagus is divided into three distinctive structural and functional zones (Fig. 1) [19]. The most superficial zone, the stratum granulosum, is four to five cell layers thick [19]. Epithelial cells in this layer function both as a physical and chemical barrier against the luminal environment. The physical epithelial barrier restricts the passage of fluids and solutes from the lumen into the deeper tissue layers. The chemical barrier consists of antimicrobial peptides, such as human β-defensin (hBD)-1, that have activity against gram-positive and gram-negative bacteria, fungi, and enveloped viruses [20]. The middle layer, the stratum spinosum, is composed of metabolically active cells that transport various molecules across their cell membranes [19]. Progenitor cells that proliferate to replenish cells in the other layers are found in the deepest layer, the stratum basalis.

The esophageal epithelial barrier is confined to the stratum granulosum and upper layers of the stratum spinosum [21]. The physical epithelial barrier is composed of the hydrophobic lipid bilayer of the squamous cell membranes and the plasma membrane-based cell-to-cell adhesion structures, including tight junctions, adherens junctions, and desmosomes (Fig. 1). Tight junction proteins include occludin, claudin, junctional adhesion molecule (JAM)-A, and zona occludens (ZO) family members (ZO-1, ZO-2,

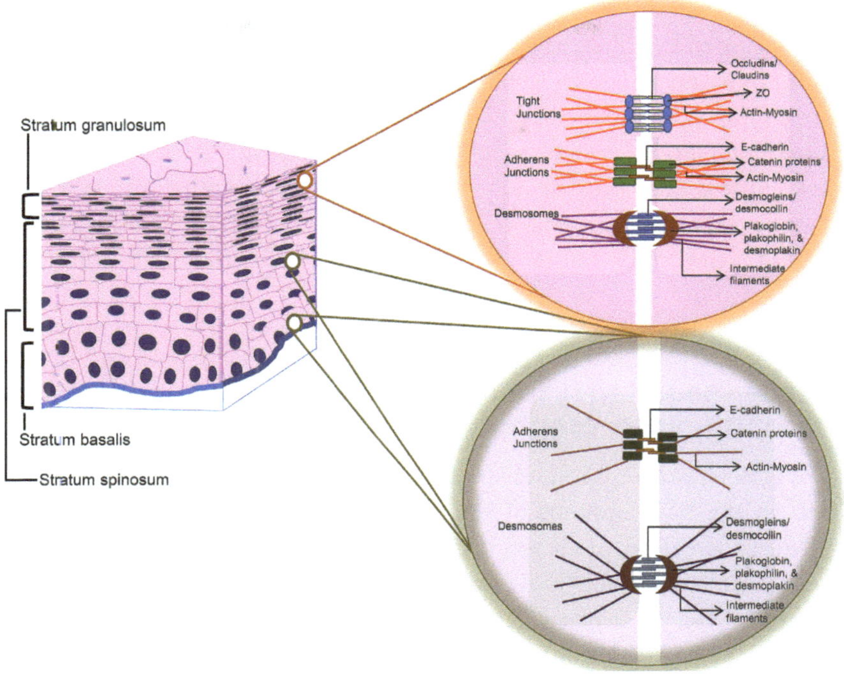

Fig. 1 Nonkeratinized, stratified squamous epithelium of the human esophagus with its three distinctive layers—stratum granulosum, stratum spinosum, and stratum basalis. The physical esophageal barrier is confined to cells in the stratum granulosum and upper layers of the stratum spinosum by the arrangement of tight and adherens junctions that encircle the cell perimeter and are anchored to a highly organized F-actin-myosin cytoskeletal network. Neighboring cells in the lower layers of the stratum spinosum and the stratum basalis are outfitted with "spot welds" that are formed by desmosomes, which are anchored to intermediate filaments, and adherens junctions, which are associated with a less organized F-actin-myosin cytoskeletal network. These lower layers do not form an important physical epithelial barrier. Key proteins in the tight junctions, adherens junctions, and desmosomes are depicted. ZO, zona occludens. *(Image from Paris S, Ekeanyanwu R, Jiang Y, Davis D, Spechler SJ, Souza RF. Obesity and its effects on the esophageal mucosal barrier. Am. J. Physiol. Gastrointest. Liver Physiol. 2021;321(3):G335–G343, with permission.)*

and ZO-3) [22]. Adherens junction proteins include E-cadherin and nectin family members and catenin family members (p120 catenin, β-catenin, and α-catenin) [22]. In these layers, the tight junction and adherens junction complexes encircle the entire cell perimeter anchored to a highly organized cortical F-actin-myosin network, blocking the passage of luminal contents between adjacent cells [23–25]. In contrast, neighboring cells in the lower layers of the stratum spinosum and the stratum basalis are outfitted with "spot welds," formed by adherens junctions, anchored to a less organized F-actin-myosin network, and desmosomes, anchored to intermediate filaments [21]. Desmosomal proteins include cadherin family members desmogleins and desmocollins, as well as plakoglobin, plakophilins, and desmoplakin [26]. These "spot welds" provide adhesive strength to the epithelium, allowing it to resist mechanical stress [21]. Regardless of the exact mechanisms underlying RE, disruption of esophageal mucosal defense systems must be involved.

Immune-mediated response in animal models and humans with acute reflux esophagitis

In 2009, Souza et al. conducted a systematic study of the early histologic events in the development of RE using a rat model in which RE was induced by creation of an esophagoduodenostomy [9]. Three days after the induction of reflux, the esophageal surface mucosa was intact, and T lymphocytes were observed infiltrating the submucosa [9]. Over the ensuing 3 weeks, T lymphocytes increased to reach the lamina propria and the epithelial layer [9]. Lymphocytic infiltration was followed by that of neutrophils; eosinophils were rarely evident. Basal cell hyperplasia and spongiosis (i.e. dilated intercellular spaces [DIS]) of cells in the stratum basalis and lower layers of the stratum spinosum were apparent by 1 week after esophagoduodenostomy and papillary hyperplasia by 2 weeks, even though the surface epithelium remained intact without the presence of erosions [9]. Erosions were seen at week 4. This sequence of events is exactly opposite of those proposed in the caustic, direct chemical injury hypothesis of RE. In the rat model of acute RE, esophageal inflammation starts below the surface contact with T lymphocytes infiltrating the submucosa with subsequent progression to the luminal surface followed by surface cell death. This is in contrast to the caustic, chemical injury model where esophageal inflammation starts at the luminal surface contact with consequent cell death, granulocyte infiltration, and subsequent progression into the submucosa.

To explore whether this histologic sequence of events in rats with acute RE was applicable to humans, Dunbar et al. induced acute RE by interrupting proton pump inhibitor (PPI) therapy for 2 weeks in 12 patients with a history of Los Angeles Grade C RE that had been healed by PPIs [10]. Within 2 weeks off PPIs, all patients had redeveloped RE documented by endoscopy as well as other parameters of gastroesophageal reflux disease (GERD) evident by 24-h pH-impedance monitoring and GERD-Health-Related Quality of Life (HRQL) symptom scores. Mucosal biopsies taken from areas in the distal esophagus that had no visible erosions demonstrated significant increases in epithelial infiltration of almost exclusively T lymphocytes by 1 and 2 weeks off PPIs. Only occasional neutrophils and eosinophils were observed. Basal cell and papillary hyperplasia also developed in areas without surface erosions, and DIS were observed in the stratum basalis and lower layers of the stratum spinosum [10]. These observations in patients with acute RE were consistent with those in the rat model of acute RE and suggest that RE develops as an immune-mediated injury, a finding that has challenged the decades' old dogma of a direct, topically caustic, chemical injury pathogenesis.

Cytokines: Initiators of the immune-mediated response in acute reflux esophagitis

Cytokines are a broad category of proteins that are produced and secreted by a wide range of cells, including immune cells, fibroblasts, endothelial cells, and epithelial cells. Chemokines, also called chemotactic cytokines, act as chemoattractants to guide cell migration along a concentration gradient toward the source of the chemokine production [27]. Because the initial observation in the rat and human models of acute RE was of immune cell infiltration initially into the submucosa with subsequent progression into the epithelium, the investigators postulated that the source of inflammatory cytokines (including chemokines) was the reflux-exposed esophageal squamous cells. Indeed, expression of interleukin (IL)-8, a pro-inflammatory cytokine (chemokine), was detected in the upper third of the squamous epithelium in rats with surgically induced RE [9]. Moreover, esophageal mucosal biopsies from patients with nonerosive esophagitis demonstrate increased levels of IL-8, suggesting that gastroesophageal reflux may stimulate IL-8 secretion without a caustic injury [28] However, rat and human esophageal mucosa contain not only squamous epithelial cells, but also inflammatory and stromal cells that could contribute to pro-inflammatory cytokines production in RE.

To determine whether the squamous cells were directly responsible for reflux-induced cytokine secretion, Souza et al. performed accompanying experiments using *in vitro* cultures of esophageal squamous cells derived from patients with GERD. [9] Exposure to solutions of acidified bile salts induced these cultured squamous cells to secrete a number of potent, pro-inflammatory cytokines, including IL-8, and conditioned media from these cells induced migration of lymphocytes and neutrophils, supporting the concept that inflammatory cytokine secretion by esophageal squamous cells led to the inflammatory infiltrate observed in the esophageal epithelium during acute RE. In another rat model of severe RE, Yamaguchi et al. found an increase in esophageal mucosal expression of inflammatory cytokines, including IL-8 within 3 h after the surgical induction of reflux, followed by infiltration of numerous immune cells, including neutrophils [29]. Administration of an antibody that blocked neutrophil infiltration of the tissue prevented some manifestations of RE, including surface epithelial cell injury [29]. Thus, these findings suggest that acute RE develops as an immune-mediated injury initiated by the secretion of cytokines from reflux-exposed esophageal squamous cells, and that the infiltrating inflammatory cells contribute to surface epithelial cell injury.

Cytokines: Initiators of epithelial repair responses in acute reflux esophagitis

In addition to attracting immune cells to the epithelium, it is also plausible that reflux-induced secretion of pro-proliferative cytokines by the epithelial cells and/or the infiltrating immune cells could stimulate reparative responses by the epithelial cells. Indeed, biopsy tissues from patients with nonerosive esophagitis have shown a correlation between high levels of IL-8 expression, the presence of intraepithelial neutrophils, and basal cell hyperplasia [28]. Moreover, RE can lead to disruption (i.e. erosion and ulcerations) of esophageal surface cells, which can be repaired by restitution (i.e. migration of cells adjacent to the wound) or by regeneration (i.e. proliferation of progenitor cells). Roudebush et al. found that esophageal squamous cells in culture chronically exposed to strongly acidic bile salt solutions secreted Activin A, a cytokine that participates in inflammatory and tissue repair processes [30]. In fact, Activin A was found to increase squamous cell migration on a basement membrane matrix, an *in vitro* model of epithelial restitution [30]. Moreover, Activin A has been shown to maintain the expression of stem cell factors and to promote stem cell self-renewal in adult tissues [31–34]. In those cultured esophageal squamous cells chronically exposed to strongly acidic bile salt solutions, Activin A was also found to increase the expression of stem cell markers, including markers of esophageal squamous progenitors and interestingly, markers of de-differentiation suggestive of reactivation of developmental, embryonic pathways [30,35]. As the embryonic esophagus is lined by ciliated, simple columnar cells, Activin A-induced de-differentiation of adult squamous esophageal stem cells to embryonic precursors may provide progenitor cells for the development of Barrett's esophagus (BE).

Cytokines: Potential disruptors of the physical epithelial barrier in acute reflux esophagitis

Inflammatory cytokines produced by reflux-exposed esophageal epithelial cells may have direct effects on the physical esophageal epithelial barrier. Patients with RE often have dilated intracellular spaces (DIS), defined as >1.5 µm by transmission electron microscopy. This is used as an index of esophageal mucosal permeability [36]. In fact, it has been proposed that finding DIS in esophageal mucosal biopsies is the earliest indicator of reflux disease [4,36]. However, studies using rabbit esophageal mucosa have found that short, physiologic exposures to strongly acidic (pH 2.0) and weakly acidic (pH 5.0) bile salt solutions increase esophageal mucosal permeability even in the absence of DIS [37]. As previously discussed, the esophageal epithelial barrier is confined to cells in the stratum granulosum and upper layers of the stratum spinosum because of the organization of tight junction and adherens junction complexes that cause a tight (i.e. low permeability) barrier. Compared with healthy controls, esophageal biopsy specimens from patients with RE show a decrease in expression of tight junction proteins claudin-1 and -4, and the presence of a cleaved, inactive form of E-cadherin, an adherens junction protein [3,38,39]. These findings have been traditionally interpreted as supporting an acid-peptic attack on cell junction proteins resulting in an increase in mucosal permeability. However, pro-inflammatory cytokines themselves have been reported to disrupt the structure and function of tight junctions and adherens junctions by decreasing expression of junctional proteins, altering trafficking of junctional proteins, inducing actin-myosin cytoskeletal abnormalities, and disassembling junctional complexes [24,25]. Thus, it is conceivable that cytokines secreted by reflux-exposed epithelial cells and/or the infiltrating immune cells could disrupt the physical esophageal barrier resulting in an increase in mucosal permeability [9].

Hypoxia-inducible factor-2α: Central role in the squamous cell cytokine response in acute reflux esophagitis

As previously discussed, blood supplied to the esophagus increases in response to luminal acid exposure as part of its post-epithelial mucosal defense system. Indeed, confocal laser endomicroscopy images in patients with acute RE demonstrate significant increases in capillary width, observations consistent with vascular dilation to increase esophageal epithelial blood supply [10]. Because a fundamental purpose of increasing blood supply is to deliver oxygen to the tissue, these findings suggest that acute RE generates a local hypoxic microenvironment. Indeed, inflammatory processes have been found to result in profound tissue hypoxia [40].

Cellular adaptation to hypoxia is mediated by the induction of hypoxia inducible factors (HIFs). HIFs are heterodimers of α subunits (HIF-1, HIF-2, or HIF-3), which are oxygen regulated, joined to a constitutively expressed β subunit [41]. These HIF heterodimers are transcription factors that drive the expression of genes that respond to hypoxic stress, and HIFs are known to mediate some inflammatory processes [41–44]. The HIF α subunits contain a highly conserved, oxygen-dependent degradation domain, which under normal oxygen conditions is hydroxylated by prolyl hydroxylases (PHDs), initiating the rapid degradation of the HIF-α subunit by proteasomes. Hypoxia decreases PHD activity, which stabilizes the HIF-α subunit allowing it to bind with the HIF-β subunit. This heterodimer then translocates to the nucleus to induce transcription of target genes, which include a number of pro-inflammatory cytokines and modulators of vascular tone [45–47]. The same as seen with hypoxia, reactive oxygen species (ROS) also can decrease PHD activity and stabilize HIF proteins, and esophageal squamous cells exposed to acidic bile salt solutions *in vitro* have been shown to produce ROS via nicotinamide adenine dinucleotide phosphate (NADPH) oxidase [48]. Based on these observations, it is conceivable that esophageal squamous cells exposed to gastric refluxate produce ROS that inactivate PHDs, enabling the formation of HIF heterodimers and the production of pro-inflammatory molecules (Fig. 2) [49].

Using esophageal biopsy specimens at 1 and 2 weeks after stopping PPIs in the aforementioned clinical study by Dunbar et al., acute RE was

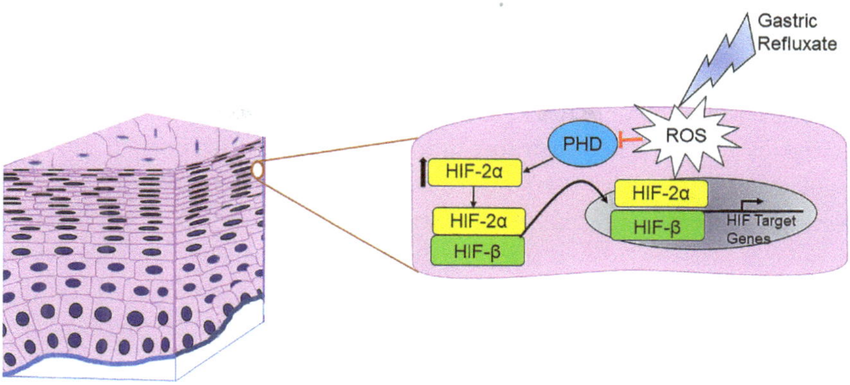

Fig. 2 Proposed mechanism of HIF-2α activation in reflux esophagitis. Esophageal squamous epithelial cells exposed to gastric refluxate generate ROS via NAPDH oxidase. ROS inhibit the activity of PHD, the enzyme that initiates rapid degradation of the HIF-α subunit by proteasomes, thus stabilizing HIF-2α in the cytoplasm. HIF-2α then binds to HIF-β and translocates to the nucleus to induce transcription of HIF target genes, including pro-inflammatory molecules. ROS, reactive oxygen species; PHD, prolyl hydroxylase; HIF, hypoxia inducible factor.

associated with a significant increase in squamous epithelial immunostaining for HIF-2α, but not HIF-1α [49].

Furthermore, acute RE was associated with increased mRNA expression of a number of HIF-2α pro-inflammatory target molecules, including IL-8, IL-1β, tumor necrosis factor-α, and cyclooxygenase-2 [49]. A series of *in vitro* experiments using cultures of human esophageal squamous cells showed that: (1) acidic bile salts generated ROS via the squamous cell's NADPH oxidase system, (2) increased ROS caused a decrease in PHD function leading to sustained increases in HIF-2α levels, (3) increased HIF-2α-mediated nuclear factor kappa-light-chain-enhancer of activated B cells (NF-κB)/p65-dependent increases in pro-inflammatory mRNA expression, and (4) inhibition of HIF-2α (by genetic shRNA knockdown or a selective small molecule antagonist) blocked acidic bile salt-mediated increases in pro-inflammatory molecule expression. In addition, inhibition of HIF-2α also blocked the acidic bile salt-induced increase in T lymphocyte migration, indicating that esophageal squamous cells secrete chemokines that attract T lymphocytes into the tissue through a HIF-2α-dependent mechanism. These observations suggest that HIF-2α plays a central role in the squamous epithelial cell inflammatory cytokine response during acute RE.

Role of the immune system in reflux esophagitis

Although, in general, immune responses are beneficial, they do come with consequences, including marked tissue damage requiring tissue regeneration for repair. Thus, it is not surprising that the immune system is subject to tight regulation [50]. Immune responses can be divided into innate and adaptive immunity. Innate immunity is an older, evolutionarily conserved, first line of defense strategy against invading pathogens and tissue damage, events which generally occur at barrier sites such as the esophagus [51,52]. Esophageal squamous epithelial cells provide both a physical and chemical (i.e. hBD-1) barrier and are the first line of defense against the luminal environment. Thus, it should not be surprising that epithelial cells are an integral component of innate immunity. Cells under attack release chemically mediated "danger signals" that affect other cells of the innate immune system, including epithelial cells, stromal cells, and fibroblasts, as well as immune cells residing in the lamina propria such as macrophages, dendritic cells, and innate lymphoid cells. Detection of danger signals by these innate immune cells leads to the generation of a local inflammatory response via the secretion of cytokines and chemokines. These in turn, recruit neutrophils, monocytes, and additional innate immune cells from the circulation into the tissue. Hypoxia is a common feature of the microenvironment produced by activation of the innate immune system, which likely contributes to disease progression [40]. Cells of the innate immune system also present antigens (generated by infected or damaged cells) to T- and B-lymphocytes, effector cells of the adaptive immune response, effecting specific and long-term memory responses [52]. Thus, the innate and adaptive immune systems are interlinked, providing a highly integrated defense system against foreign invasion. Despite recognition over 10 years ago that acute RE may be caused by a cytokine-induced immune-mediated injury, very little is known about the healthy esophageal immune system and modulation during RE.

Esophageal innate immunity and toll-like receptors

Innate immune responses are triggered by a series of germline-encoded receptors termed pattern recognition receptors (PRRs) that can recognize "danger signals" in the form of pathogen-associated molecular patterns (PAMPs) and damage-associated molecular patterns (DAMPs) which are released by stressed, injured, or dying cells [53]. Upon ligand recognition, PRRs activate downstream signaling cascades which produce type

I interferon, inflammatory cytokines, and a number of other inflammatory mediators. In addition to protection from microbial invasion, the cellular release of mediators upon PRR activation also governs key processes, including cell proliferation, cell death, and epithelial barrier integrity [54]. The most common families of PRRs include toll-like receptors (TLRs), retinoic acid-inducible gene I-like receptors, nucleotide oligomerization domain-like receptors (NLRs), and cytosolic DNA sensors [53]. Specificity of the PRRs is determined either by their ligands or cellular location.

In humans, the TLR family contains 10 members—TLR1-TRL10. All TLRs have an extracellular domain containing leucine-rich repeats that mediate ligand specificity, a transmembrane domain, and a cytoplasmic toll-IL-1 receptor domain that triggers downstream signaling cascades resulting in activation of NF-κB, interferon regulatory factors or mitogen-activated protein kinases to regulate the expression of cytokines, chemokines, type I interferons, and type III interferons [55–57]. TLRs can be expressed at the cell surface or in cytoplasmic compartments such as the endoplasmic reticulum, endosomes, and lysosomes. Surface receptors (TLR 1, 2, 4, 5, 6, and 10) respond to bacterial ligands such as lipids, lipoproteins, and proteins, whereas intracellular receptors (TLR 3, 7, 8, and 9) recognize bacterial and viral nucleic acids [57]. In esophageal mucosa, squamous epithelial cells, myofibroblasts cells, fibroblasts, endothelial cells, and inflammatory cells in the lamina propria have been found to express TLRs [14,58].

Primary esophageal squamous epithelial cells from pediatric and adult patients and in vitro cultures of telomerase-immortalized, esophageal squamous cell lines demonstrate mRNA expression of TLR1 through TLR6, with minimal mRNA expression for TLR7 through TLR10 [59,60]. Compared with esophageal biopsies from normal controls, those from adult patients with RE demonstrate increased mRNA expression of TLR2 and TLR4 [58,61]. Verbeek et al. used immunohistochemistry to document the location of TLR2- or TLR4-expressing cells in esophageal mucosal biopsy specimens [58,61]. In normal control biopsies, TLR2 and TLR4 proteins are found in the nucleus and cytoplasm of squamous cells located in the stratum spinosum and the stratum basalis, including those basal cells that are adjacent to the lamina propria papillae; inflammatory cells in the lamina propria were also found to express TLR2 and TLR4 proteins. In biopsy specimens from patients with RE, TLR2- and TLR4-expressing cells were found in the stratum spinosum, stratum basalis, and alongside the papillae, with both proteins expressed by squamous cells in the more superficial epithelial layers [58,61,62]. It remains unclear however, as to why cells in the more superficial layers express TLR2 and TLR4 in RE. Esophageal myofibroblasts have been found to express mRNAs for TLR2 and TLR4, with TLR4 mRNA demonstrating a 30-fold increase compared with TLR2 [14]. TLR4 protein has been detected at the surface and in the cytoplasm of esophageal myofibroblasts. Furthermore, in biopsy specimens from patients with RE, the myofibroblast population was found to significantly increase when compared with that of the normal esophagus [14].

Activation of the innate immune system during reflux esophagitis

Esophageal epithelial cells at the luminal surface provide first-line immune response to invading pathogens. The esophageal mucosal surface is covered by a thin layer of microbiota, in contrast to the extremely dense microbial community that covers the surfaces of the intestine and colon [55,63]. The normal microbiome does not cause a pro-inflammatory response. It is only when immune defense mechanisms are impaired or if microbes break through the mucosal barrier that the immune system reacts. It has been shown that intestinal epithelial cells express TLRs only at their basolateral surface, an area that is not normally in contact with luminal microbes. In fact, it has been proposed that the basolateral expression of TLRs by intestinal cells is an adaptation to their extremely dense microbial community to prevent inappropriate receptor activation, which can lead to uncontrolled chronic inflammation and promote carcinogenesis [64–66]. In healthy esophageal epithelium, TLRs are expressed on squamous cells that reside in the deeper epithelial layers and on cells in the lamina propria. Similar to the intestine, the expression of TLR in these deeper locations is likely an adaptation to the harsh luminal environment encountered by cells at the esophageal surface to prevent uncontrolled chronic inflammation. Therefore, activation of esophageal TLRs would require a breach in the epithelial barrier. Although very little is known about the esophageal chemical barrier, Schroeder et al. found there was no significant difference in expression of human-β defensin-1 or human-β defensin-3 between esophageal tissues from healthy controls and patients with reflux disease [20]. Therefore, it is likely that the breach occurs in the physical (i.e. structural) barrier during RE, perhaps by cytokine-induced effects on tight junction and adherens junction complexes.

TLR2 signaling in esophageal squamous mucosa

TLR2 is a cell surface receptor that recognizes a wide range of ligands, including gram-negative and gram-positive bacterial products and ligands released by stressed tissue, including heat shock proteins and high mobility group box 1 (HMGB1) protein [67,68]. The healthy esophageal microbiome consists primarily of gram-positive microbes, whereas the microbiome in patients with RE is dominated by gram-negative bacteria [69–71]. In addition, esophageal squamous cell stress, damage, and necrosis can occur during RE [4,9,30], resulting in the release of DAMPs by these injured or dying cells that could activate TLR2 on surrounding cells [72–74]. Depending on cell type and ligand, TLR2 activation may lead to pro-inflammatory or antiinflammatory responses. Ruffner et al. explored the role of TLR2 activation on epithelial barrier permeability using esophageal squamous cells in air–liquid interface (ALI) cultures which allow for epithelial stratification and differentiation [59,75]. In this study, TLR2 agonists were added to the media in the basolateral chamber of the ALI culture adjacent to the basal-like cells. TLR2 signaling was found to "tighten" epithelial barrier permeability, which was associated with increases in expression of tight junction proteins ZO-1 and claudin-1. These findings suggest that TLR2 signaling by cells located in the deeper layers of the esophagus might play a critical role in restoring barrier integrity during RE.

TLR4 signaling in esophageal squamous mucosa

TLR4 is a cell surface receptor known primarily for its ability to bind lipopolysaccharide (LPS), a component of the outer membrane of gram-negative bacteria, but also binds to host-derived DAMPs. Following ligand binding, TLR4 activates NF-kB and its downstream target genes, including those involved in inflammation, chemoattraction, proliferation, and antimicrobial defense. Controversy exists regarding the role of TLR4 activation in esophageal squamous cells, however. Verbeek et al. reported that LPS caused a small, but statistically significant increase in IL-8 secretion in esophageal squamous epithelial biopsy tissues, whereas Nadatani et al. did not find LPS-induced increases in IL-8 or TNF-α in esophageal squamous epithelial cells in culture [58,76]. It is conceivable that the observations by Verbeek et al. stemmed from the use of squamous biopsy specimens that may have included inflammatory and stromal cells in addition to epithelial cells.

Those nonepithelial cells also express TLR4, and the small increases in LPS-induced IL8 secretion may have resulted from TLR4 signaling in those cells, rather than from squamous epithelial cells. In support of this notion, Gargus et al. reported that esophageal myofibroblasts secrete IL-6 and IL-8 in response to TLR4 activation by LPS or by HMGB1 [14]. Using mouse models in which RE was induced by a surgical duodeno-gastroesophageal anastomosis (DGEA), Gergen et al. explored the role of TLR4 in reflux-induced proliferative changes of the epithelium [77]. CH3/HeJ mice demonstrate a spontaneous point mutation in TLR4 ($TLR4^{Lps-d}$) leading to inefficient receptor binding of LPS. Six weeks after DGEA, TLR4 mutant mice displayed significant reductions in esophageal squamous mucosal thickening and epithelial proliferation compared with wild-type controls, suggesting that LPS-mediated TLR4 signaling contributes to reflux-induced proliferative epithelial responses [77]. In addition, cultures of esophageal myofibroblasts exposed to acidic bile salts in vitro have been found to secrete factors which increase epithelial basal cell proliferation via paracrine interactions [78]. Thus, it is conceivable that nonepithelial cell signaling via TLR4 might also play a critical role in restoring epithelial barrier integrity during RE.

Esophageal adaptive immunity and T cell phenotypes

As discussed above, T lymphocytes are among the effector cells of the adaptive immune responses. T cell progenitors originate in the bone marrow or fetal liver and enter the thymus where they undergo a series of positive and negative selection processes before committing to either the CD4+ helper or CD8+ cytotoxic cell lineages; regulatory and natural killer T cells are also generated in the thymus [79]. Following an antigen encounter, naïve CD4+ T cells differentiate into a number of effector T cell subgroups, including T helper 1 (T_H1) and T_H2, which express unique cytokine profiles, and naïve CD8+ T cells differentiate into cytotoxic T cells, which produce cytotoxic enzymes to kill pathogen-infected host cells. Following antigen elimination, activated CD4+ and CD8+ T cells generate memory T cells allowing for an enhanced immune response when the antigen is once again encountered.

Rat and human models of acute RE have demonstrated that T cells are the first immune cells to infiltrate the reflux-damaged esophagus [10]. In esophageal biopsy, tissues from patients with RE, elevated mRNA levels of IL-1β and IL-8 have been reported [17,28,80,81]. Fitzgerald et al.

performed immunohistochemical staining for IFN-γ, IL-1β, and IL-8 in RE tissues to gain insight into the cellular source for production of these cytokines. They found that IFN-γ was exclusively produced by T cells in the lamina propria, whereas IL-1β and IL-8 were produced by both epithelial cells and T cells, suggesting that the T cell population in RE exhibits a pro-inflammatory T_H1 cytokine profile [17,79]. A few studies have found resident CD8+ T lymphocytes in healthy esophageal squamous epithelium [16,82]. Lind et al. have expanded these data by further characterizing the T cell phenotype by isolating esophageal tissue lymphocytes from mucosal biopsies of patients with RE and normal control subjects, expanding those lymphocytes ex vivo, and immunophenotyping them by fluorescence activated cell sorting (FACS) analysis [15]. Biopsies from RE patients demonstrated a significantly higher frequency of CD4+ cells with lower numbers expressing CD103+ compared with those isolated from biopsies of normal controls. CD103 can be upregulated on T cells by TGF-β, which is usually present in the tissue and, therefore, T cells dwelling within the tissue tend to be CD103+. Furthermore, CD103+ lymphocytes tend to have an immunosuppressive phenotype [83,84]. Therefore, low numbers of CD4+CD103+ T cells in RE biopsies not only support a pro-inflammatory environment but also suggest that the predominant CD4+CD103- T cell population was likely trafficked from the circulation into the tissue [15]. Biopsies from both RE and controls had a high frequency of CD8+CD103+ cells, but those from RE patients had significantly higher levels of CD8+GranzymeB+ – cells [15,16]. Granzyme B positivity is an indicator of cytotoxic CD8+ effector cells [85]. These CD8+GranzymeB+ cells were both CD103+ and CD103-, suggesting that both resident (CD8+CD103+) and recruited (CD8+CD103-) T cells are stimulated to become cytotoxic by the tissue microenvironment of RE [15].

Conclusions

For over 80 years, the pathogenesis of RE has been assumed to be a caustic chemical injury (i.e., an acid burn). More recent evidence now suggests that the pathogenesis of RE is initiated by an immune-mediated injury (Fig. 3). In this model, the process starts with the release of cytokines by reflux-exposed esophageal squamous cells via a HIF-2α-dependent mechanism. These cytokines attract T lymphocytes, with a pro-inflammatory T_H1 cytokine profile and with cytotoxic features, from the circulation into the esophageal submucosa, with subsequent progression to the luminal surface.

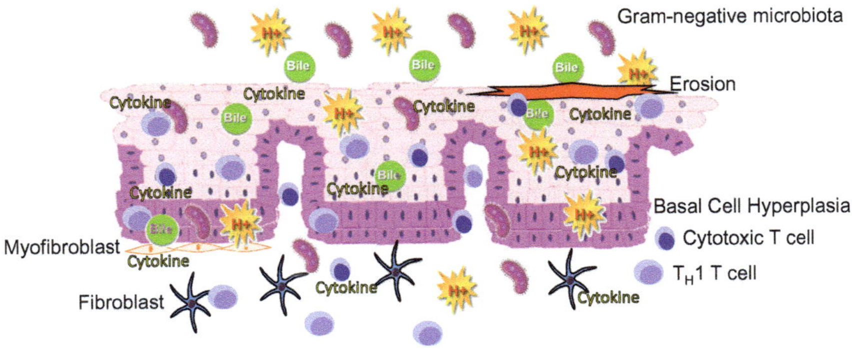

Fig. 3 Conceptual Model for the Cytokine-Induced, Immune-Mediated Pathogenesis of Reflux Esophagitis. Esophageal squamous cells exposed to acid and bile salts, the main noxious constituents of gastric refluxate, release pro-inflammatory cytokines that recruit T_H1 and cytotoxic T cells from the circulation into the submucosa of the tissue with subsequent progression to the luminal surface. Pro-inflammatory cytokines disrupt the structure and function of tight junctions and adherens junctions that form the physical epithelial barrier, allowing the penetration of luminal contents such as acid, bile salts, and the reflux-associated, gram-negative esophageal microbiota into the deeper layers of the epithelium. These luminal contents and products released by reflux-damaged surface squamous cells activate TLRs found on squamous cells in the deeper layers of the epithelium and on cells in the lamina propria, including subepithelial myofibroblasts, fibroblasts, and immune cells, further leading to production of pro-inflammatory cytokines, chemokines, and interferons. Moreover, release of pro-proliferative cytokines by epithelial cells, myofibroblasts, and immune cells can lead to basal cell hyperplasia. As reflux continues, mucosal integrity is disrupted leading to the formation of erosions and ulcerations, wounds that undergo cytokine-mediated tissue repair and regeneration. Thus, recent data suggest that reflux esophagitis begins as a cytokine-sizzle rather than as an acid burn.

Pro-inflammatory cytokines released by epithelial cells and T cells can disrupt the structure and function of tight junction and adherens junction complexes that form the epithelial permeability barrier, allowing the penetration of luminal contents such as acid, bile salts, and the reflux-associated, gram-negative esophageal microbiota into the deeper layers of the epithelium. Microbial contents such as LPS and products released by reflux-damaged surface squamous cells can then activate TLRs found on squamous cells in the deeper layer of the epithelium, subepithelial myofibroblasts, and immune cells in the lamina propria, leading further to pro-inflammatory cytokines, chemokines, and interferon production. Moreover, release of pro-proliferative cytokines by epithelial cells, myofibroblasts, and immune cells can lead to basal cell hyperplasia as a reparative response. As reflux continues, mucosal integrity is disrupted leading to the formation of erosions and ulcerations, injuries which can be repaired by cytokine-mediated tissue repair and regenerative processes. Thus, recent literature suggests that mucosal RE begins as a *bottom-up* directed "cytokine-sizzle," rather than a *top-down* "acid burn."

Future directions

A clearer understanding of the cells and mechanisms involved in the development of RE is needed to develop effective strategies for preventing and treating this inflammatory complication of gastroesophageal reflux disease. Furthermore, avoiding reflux-induced tissue damage that requires tissue regeneration for repair might prevent BE and esophageal adenocarcinoma. Increased mucosal permeability appears to be a key event in the amplification of inflammatory signaling by allowing toxic luminal contents access to cells in the deeper layers of the epithelium and lamina propria. Uncovering mechanisms of mucosal permeability and defining the cellular landscape of RE will identify targets for diagnostic or therapeutic utility in the future. Thus, future directions should include:

- Studying mechanisms whereby RE increases mucosal permeability.
- Examining whether changes in the esophageal microbiome contribute to or result from reflux-induced mucosal inflammation.
- Characterization of additional components of esophageal innate immunity such as endogenous antimicrobials and their function in esophageal homeostasis and reflux disease.
- Exploration into the role of crosstalk between epithelial cells and microbes, epithelial cells and myofibroblasts, and epithelial cells and other stromal cells to identify molecules that may influence mucosal permeability.
- Applying newer technologies such as single-cell transcriptomics or genomics to esophageal biopsies obtained from patients with acute and chronic RE, both on and off PPI therapy, to allow for identification of gene signatures that single out epithelial and nonepithelial cell contributions to disease initiation, progression, and resolution.
- Discovering gene signatures of esophageal stem or progenitor cells during RE to map the routes toward epithelial regeneration either by squamous cells or by metaplastic BE cells.

COI disclosures

Dr. Souza receives research support from Phathom Pharmaceuticals and Sanofi and is a consultant for Ironwood Pharmaceuticals, Castle Biosciences, Phathom Pharmaceuticals, IsoThrive, CDx, and AstraZeneca.

References

[1] A. Winkelstein, Peptic esophagitis: a new clinical entity, JAMA 104 (1935) 906–909.
[2] N.A. Tobey, S.S. Hosseini, C. Caymaz-Bor, H.R. Wyatt, G.S. Orlando, R.C. Orlando, The role of pepsin in acid injury to esophageal epithelium, Am. J. Gastroenterol. 2001 (96) (2001) 3062–3070.
[3] B. Jovov, J. Que, N.A. Tobey, Z. Djukic, B.L. Hogan, R.C. Orlando, Role of E-cadherin in the pathogenesis of gastroesophageal reflux disease, Am. J. Gastroenterol. 106 (2011) 1039–1047.
[4] R.C. Orlando, D.W. Powell, C.N. Carney, Pathophysiology of acute acid injury in rabbit esophageal epithelium, J. Clin. Invest. 68 (1) (1981) 286–293.
[5] R.C. Orlando, Pathogenesis of reflux esophagitis and Barrett's esophagus, Med. Clin. NA 89 (2005) 219–241.
[6] H.I. Goldberg, W.J. Dodds, S. Gee, C. Montgomery, F.F. Zboralske, Role of acid and pepsin in acute experimental esophagitis, Gastroenterology 56 (1969) 223–230.
[7] R. Fiocca, L. Mastracci, R. Riddell, K. Takubo, et al., Development of consensus guidelines for the histologic recognition of microscopic esophagitis in patients with gastroesophageal reflux disease: the Esohisto project, Hum. Pathol. 41 (2010) 223–231.
[8] F. Ismail-Beigi, P.F. Horton, C.E. Pope 2nd, Histological consequences of gastroesophageal reflux in man, Gastroenterology 58 (1970) 163–174.
[9] R.F. Souza, X. Huo, V. Mittal, C.M. Schuler, et al., Gastroesophageal reflux might cause esophagitis through a cytokine-mediated mechanism rather than caustic acid injury, Gastroenterology 137 (2009) 1776–1784.
[10] K.B. Dunbar, A.T. Agoston, R.D. Odze, et al., Association of acute gastroesophageal reflux disease with esophageal histologic changes, JAMA 315 (2016) 2104–2112.
[11] R.C. Orlando, Review article: oesophageal mucosal resistance, Aliment. Pharmacol. Ther. 12 (1998) 191–197.
[12] B.L. Bass, E.J. Schweitzer, J.W. Harmon, J. Kraimer, H+ back diffusion interferes with intrinsic reactive regulation of esophageal mucosal blood flow, Surgery 96 (1984) 404–413.
[13] M.E. Hollwarth, M. Smith, P.R. Kvietys, D.N. Granger, Esophageal blood flow in the cat. Normal distribution and effects of acid perfusion, Gastroenterology 90 (1986) 622–627.
[14] M. Gargus, C. Niu, J.G. Vallone, J. Binkley, D.C. Rubin, A. Shaker, Human esophageal myofibroblasts secrete proinflammatory cytokines in response to acid and Toll-like receptor 4 ligands, Am. J. Physiol. Gastrointest. Liver Physiol. 308 (2015), G904–923.
[15] A. Lind, L. Koenderman, J.G. Kusters, P.D. Siersema, Squamous tissue lymphocytes in the esophagus of controls and patients with reflux esophagitis and Barrett's esophagus are characterized by a non-inflammatory phenotype, PLoS One 9 (2014), e106261.
[16] M.B. Resnick, Y. Finkelstein, A. Weissler, J. Levy, E. Yakirevich, Assessment and diagnostic utility of the cytotoxic T-lymphocyte phenotype using the specific markers granzyme-B and TIA-1 in esophageal mucosal biopsies, Hum. Pathol. 30 (1999) 397–402.

[17] R.C. Fitzgerald, B.A. Onwuegbusi, M. Bajaj-Elliott, I.T. Saeed, W.R. Burnham, M.J. Farthing, Diversity in the oesophageal phenotypic response to gastro-oesophageal reflux: immunological determinants, Gut 50 (2002) 451–459.
[18] C. Zhang, C. Niu, K. Yang, A. Shaker, Human esophageal myofibroblast secretion of bone morphogenetic proteins and GREMLIN1 and paracrine regulation of squamous epithelial growth, Sci. Rep. 8 (2018) 12354.
[19] X. Zhang, D. Patil, R.D. Odze, et al., The microscopic anatomy of the esophagus including the individual layers, specialized tissues, and unique components and their responses to injury, Ann. N. Y. Acad. Sci. 1434 (2018) 304–318.
[20] S. Schroeder, Z.D. Robinson, J.C. Masterson, et al., Esophageal human β-defensin expression in eosinophilic esophagitis, Pediatr. Res. 73 (2013) 647–654.
[21] M. Rubsam, J.A. Broussard, S.A. Wickstrom, O. Nekrasova, K.J. Green, C.M. Niessen, Adherens junctions and desmosomes coordinate mechanics and signaling to orchestrate tissue morphogenesis and function: an evolutionary perspective, Cold Spring Harb. Perspect. Biol. 10 (11) (2018). a029207.
[22] A. Hartsock, W.J. Nelson, Adherens and tight junctions: structure, function and connections to the actin cytoskeleton, Biochim. Biophys. Acta 1778 (2008) 660–669.
[23] T. Oshima, H. Miwa, Gastrointestinal mucosal barrier function and diseases, J. Gastroenterol. 51 (2016) 768–778.
[24] C.T. Capaldo, A. Nusrat, Cytokine regulation of tight junctions, Biochim. Biophys. Acta 1788 (2009) 864–871.
[25] S. Lechuga, A.I. Ivanov, Disruption of the epithelial barrier during intestinal inflammation: Quest for new molecules and mechanisms, Biochim. Biophys. Acta Mol. Cell Res. 1864 (7) (2017) 1183–1194.
[26] A.P. Kowalczyk, K.J. Green, Structure, function, and regulation of desmosomes, Prog. Mol. Biol. Transl. Sci. 116 (2013) 95–118.
[27] D. Raman, T. Sobolik-Delmaire, A. Richmond, Chemokines in health and disease, Exp. Cell Res. 317 (2011) 575–589.
[28] H. Isomoto, V.A. Saenko, Y. Kanazawa, Y. Nishi, A. Ohtsuru, K. Inoue, Y. Akazawa, F. Takeshima, K. Omagari, M. Miyazaki, Y. Mizuta, I. Murata, S. Yamashita, S. Kohno, Enhanced expression of interleukin-8 and activation of nuclear factor kappa-B in endoscopy-negative gastroesophageal reflux disease, Am. J. Gastroenterol. 99 (2004) 589–597.
[29] T. Yamaguchi, N. Yoshida, N. Tomatsuri, et al., T. Cytokine-induced neutrophil accumulation in the pathogenesis of acute reflux esophagitis in rats, Int. J. Mol. Med. 16 (2005) 71–77.
[30] C. Roudebush, A. Catala-Valentin, T. Andl, G.F. Le Bras, C.D. Andl, Activin A-mediated epithelial de-differentiation contributes to injury repair in an in vitro gastrointestinal reflux model, Cytokine 123 (2019), 154782.
[31] L. Vallier, S. Mendjan, S. Brown, et al., Activin/Nodal signalling maintains pluripotency by controlling Nanog expression, Development 136 (2009) 1339–1349.
[32] S. Cambray, C. Arber, G. Little, et al., Activin induces cortical interneuron identity and differentiation in embryonic stem cell-derived telencephalic neural precursors, Nat. Commun. 3 (2012) 841.
[33] M. Kadaja, B.E. Keyes, M. Lin, et al., SOX9: a stem cell transcriptional regulator of secreted niche signaling factors, Genes Dev. 28 (2014) 328–341.
[34] L. Xiao, X. Yuan, S.J. Sharkis, Activin A maintains self-renewal and regulates fibroblast growth factor, Wnt, and bone morphogenic protein pathways in human embryonic stem cells, Stem Cells (Dayton, Ohio) 24 (2006) 1476–1486.
[35] S.G. Willet, M.A. Lewis, Z.F. Miao, et al., Regenerative proliferation of differentiated cells by mTORC1-dependent paligenosis, EMBO J. 37 (2018).

[36] N.A. Tobey, J.L. Carson, R.A. Alkiek, R.C. Orlando, Dilated intercellular spaces: a morphological feature of acid reflux-damaged human esophageal epithelium, Gastroenterology 111 (1996) 1200–1205.
[37] R. Farre, H. van Malenstein, R. De Vos, et al., Short exposure of oesophageal mucosa to bile acids, both in acidic and weakly acidic conditions, can impair mucosal integrity and provoke dilated intercellular spaces, Gut 57 (2008) 1366–1374.
[38] E.V. Bjorkman, A. Edebo, M. Oltean, A. Casselbrant, Esophageal barrier function and tight junction expression in healthy subjects and patients with gastroesophageal reflux disease: functionality of esophageal mucosa exposed to bile salt and trypsin in vitro, Scand. J. Gastroenterol. 48 (2013) 1118–1126.
[39] N.A. Tobey, T.M. Gambling, X.C. Vanegas, J.L. Carson, R.C. Orlando, Physicochemical basis for dilated intercellular spaces in non-erosive acid-damaged rabbit esophageal epithelium, Dis. Esophagus 21 (2008) 757–764.
[40] S.P. Colgan, G.T. Furuta, C.T. Taylor, Hypoxia and Innate Immunity: Keeping Up with the HIFsters, Annu. Rev. Immunol. 38 (2020) 341–363.
[41] A. Loboda, A. Jozkowicz, J. Dulak, HIF-1 and HIF-2 transcription factors—similar but not identical, Mol. Cells 29 (2010) 435–442.
[42] C.T. Taylor, Interdependent roles for hypoxia inducible factor and nuclear factor-kappaB in hypoxic inflammation, J. Physiol. 586 (2008) 4055–4059.
[43] H.K. Eltzschig, P. Carmeliet, Hypoxia and inflammation, N. Engl. J. Med. 364 (2011) 656–665.
[44] A.S. Zinkernagel, R.S. Johnson, V. Nizet, Hypoxia inducible factor (HIF) function in innate immunity and infection, J. Mol. Med. (Berl) 2007 (85) (2007) 1339–1346.
[45] Y.M. Shah, S. Ito, K. Morimura, et al., Hypoxia-inducible factor augments experimental colitis through an MIF-dependent inflammatory signaling cascade, Gastroenterology 134 (2008) 2036–2048. 2048.e2031–2033.
[46] R. Singhal, Y.M. Shah, Oxygen battle in the gut: Hypoxia and hypoxia-inducible factors in metabolic and inflammatory responses in the intestine, J. Biol. Chem. 295 (2020) 10493–10505.
[47] A. Carreau, B. El Hafny-Rahbi, A. Matejuk, C. Grillon, C. Kieda, Why is the partial oxygen pressure of human tissues a crucial parameter? Small molecules and hypoxia, J. Cell. Mol. Med. 15 (2011) 1239–1253.
[48] Feagins L.A., Zhang H.Y., Zhang X., et al.. Mechanisms of oxidant production in esophageal squamous cell and Barrett's cell lines. Am. J. Physiol. Gastrointest. Liver Physiol. 2008, 294: G411–417, 2008.
[49] X. Huo, A.T. Agoston, K.B. Dunbar, et al., Hypoxia-inducible factor-2alpha plays a role in mediating oesophagitis in GORD, Gut 66 (2017) 1542–1554.
[50] S.R. Paludan, T. Pradeu, S.L. Masters, T.H. Mogensen, Constitutive immune mechanisms: mediators of host defence and immune regulation, Nat. Rev. Immunol. 21 (2021) 137–150.
[51] E.A. Mandujano-Tinoco, E. Sultan, A. Ottolenghi, O. Gershoni-Yahalom, B. Rosental, Evolution of cellular immunity effector cells; perspective on cytotoxic and phagocytic cellular lineages, Cell 10 (8) (2021) 1853.
[52] A. Guillemin, A. Kumar, M. Wencker, E.P. Ricci, Shaping the innate immune response through post-transcriptional regulation of gene expression mediated by RNA-binding proteins, Front. Immunol. 12 (2022), 796012.
[53] M.R. Thompson, J.J. Kaminski, E.A. Kurt-Jones, K.A. Fitzgerald, Pattern recognition receptors and the innate immune response to viral infection, Viruses 3 (2021) 920–940.
[54] M. Mahapatro, L. Erkert, C. Becker, Cytokine-mediated crosstalk between immune cells and epithelial cells in the gut, Cell 10 (1) (2021) 111.
[55] M. Lotz, S. Ménard, M. Hornef, Innate immune recognition on the intestinal mucosa, Int. J. Med. Microbiol. 297 (5) (2007) 379–392.

[56] S.H. Aghamiri, K. Komlakh, M. Ghaffari, The crosstalk among TLR2, TLR4 and pathogenic pathways: a treasure trove for treatment of diabetic neuropathy, Inflammopharmacology (2022).
[57] T. Kawasaki, T. Kawai, Toll-like receptor signaling pathways, Front. Immunol. 5 (2014) 461.
[58] R.E. Verbeek, P.D. Siersema, F.J. Ten Kate, et al., Toll-like receptor 4 activation in Barrett's esophagus results in a strong increase in COX-2 expression, J. Gastroenterology 49 (2014) 1121–1134.
[59] M.A. Ruffner, L. Song, K. Maurer, et al., Toll-like receptor 2 stimulation augments esophageal barrier integrity, Allergy 74 (2019) 2449–2460.
[60] D.M. Lim, S. Narasimhan, C.Z. Michaylira, M.L. Wang, TLR3-mediated NF-{kappa}B signaling in human esophageal epithelial cells, Am. J. Physiol. Gastrointest. Liver Physiol. 297 (6) (2009) G1172–G1180.
[61] R.E. Verbeek, P.D. Siersema, F.P. Vleggaar, et al., Toll-like receptor 2 signalling and the lysosomal machinery in Barrett's esophagus, J. Gastrointestin. Liver Dis. 25 (2016) 273–282.
[62] M. Nortunen, N. Väkiparta, K. Porvari, J. Saarnio, T.J. Karttunen, H. Huhta, Pathophysiology of reflux oesophagitis: role of Toll-like receptors 2 and 4 and Farnesoid X receptor, Virchows Arch. 479 (2021) 285–293.
[63] I. Nina Hornickel, J. Kacza, A. Schnapper, et al., Demonstration of substances of innate immunity in the esophageal epithelium of domesticated mammals: Part II—Defence mechanisms, including species comparison, Acta Histochem. 113 (2011) 175–188.
[64] M. Fukata, A. Chen, A.S. Vamadevan, et al., Toll-like receptor-4 promotes the development of colitis-associated colorectal tumors, Gastroenterology 133 (2007) 1869–1881.
[65] S.C. Gribar, R.J. Anand, C.P. Sodhi, D.J. Hackam, The role of epithelial Toll-like receptor signaling in the pathogenesis of intestinal inflammation, J. Leukoc. Biol. 83 (2008) 493–498.
[66] O.P. Schären, S. Hapfelmeier, Robust microbe immune recognition in the intestinal mucosa, Genes Immun. 22 (2021) 268–275.
[67] A.G. Testro, K. Visvanathan, Toll-like receptors and their role in gastrointestinal disease, J. Gastroenterol. Hepatol. 24 (2009) 943–954.
[68] M.F. Tsan, Heat shock proteins and high mobility group box 1 protein lack cytokine function, Jf. Leuko Biol. 89 (2011) 847–853.
[69] L. Yang, X. Lu, C.W. Nossa, F. Francois, R.M. Peek, Z. Pei, Inflammation and intestinal metaplasia of the distal esophagus are associated with alterations in the microbiome, Gastroenterology 137 (2009) 588–597.
[70] S. Macfarlane, E. Furrie, G.T. Macfarlane, J.F. Dillon, Microbial colonization of the upper gastrointestinal tract in patients with Barrett's esophagus, Clin. Infect. Dis. 45 (2007) 29–38.
[71] I. Amir, F.M. Konikoff, M. Oppenheim, U. Gophna, E.E. Half, Gastric microbiota is altered in oesophagitis and Barrett's oesophagus and further modified by proton pump inhibitors, Environ. Microbiol. 16 (2014) 2905–2914.
[72] D.M. Lim, M.L. Wang, Toll-like receptor 3 signaling enables human esophageal epithelial cells to sense endogenous danger signals released by necrotic cells, Am. J. Physiol. Gastrointest. Liver Physiol. 301 (2011) G91–G99.
[73] D. Foell, H. Wittkowski, J. Roth, Mechanisms of disease: a 'DAMP' view of inflammatory arthritis, Nat. Clin. Pract. Rheumatol. 3 (2007) 382–390.
[74] A.M. Piccinini, K.S. Midwood, DAMPening inflammation by modulating TLR signalling, Mediators Inflamm. 2010 (2010) 672395.
[75] K. Kc, M.E. Rothenberg, J.D. Sherrill, In vitro model for studying esophageal epithelial differentiation and allergic inflammatory responses identifies keratin involvement in eosinophilic esophagitis, PLoS One 10 (2015), e0127755.

[76] Y. Nadatani, X. Huo, X. Zhang, et al., NOD-Like receptor protein 3 inflammasome priming and activation in Barrett's epithelial cells, Cell. Mol. Gastroenterol. Hepatol. 2016 (2) (2016) 439–453.
[77] A.K. Gergen, M.J. Jarrett, A. Li, X. Meng, A. Pratap, D.A. Fullerton, M.J. Weyant, Toll-like receptor 4 mediates reflux-induced inflammation in a murine reflux model, Semin. Thorac. Cardiovasc. Surg. 15 (2021). S1043-0679(21)00417-2.
[78] L. Hu, C. Zhang, K. Yang, M. Li, A. Shaker, Human esophageal myofibroblasts increase squamous epithelial thickness via paracrine mechanisms in an in vitro model of gastroesophageal reflux disease, PLoS One 15 (2020), e0238852.
[79] A. Dutta, H. Venkataganesh, P.E. Love, New insights into epigenetic regulation of T cell differentiation, Cell 10 (12) (2021) 3459.
[80] H. Isomoto, A. Wang, Y. Mizuta, et al., Elevated levels of chemokines in esophageal mucosa of patients with reflux esophagitis, Am. J. Gastroenterol. 98 (2003) 551–556.
[81] D.S. Oh, S.R. DeMeester, D. Vallbohmer, et al., Reduction of interleukin 8 gene expression in reflux esophagitis and Barrett's esophagus with antireflux surgery, Arch. Surg. 142 (2007) 554–559. discussion 559–560.
[82] S.W. Carmack, R.H. Lash, J.M. Gulizia, R.M. Genta, Lymphocytic disorders of the gastrointestinal tract: a review for the practicing pathologist, Adv Anat Pathol 16 (2009) 290–306.
[83] S.D. Koch, E. Uss, R.A. van Lier, I.J. ten Berge, Alloantigen-induced regulatory CD8+CD103+ T cells, Hum. Immunol. 69 (2008) 737–744.
[84] E. Uss, A.T. Rowshani, B. Hooibrink, N.M. Lardy, R.A. van Lier, I.J. ten Berge, CD103 is a marker for alloantigen-induced regulatory CD8+ T cells, J. Immunol. 177 (2006) 2775–2783.
[85] I.S. Afonina, S.P. Cullen, S.J. Martin, Cytotoxic and non-cytotoxic roles of the CTL/NK protease granzyme B, Immunol. Rev. 235 (2010) 105–116.

CHAPTER 3

Acid suppression and effects on the gastrointestinal microbiome

Heather Korus[a] and Nimish Vakil[b]
[a]Aurora Health Care, Summit, WI, United States
[b]University of Wisconsin School of Medicine and Public Health, Madison, WI, United States

Introduction

The microbiome of the human gut has become an area of interest recently because of its potential role in important normative physiologic processes and a potential relationship to various disease states, including inflammatory bowel disease (IBD), obesity, and divergent conditions such as autism. This has given rise to the study of the microbiome termed "microbiomics," which is further broken down into three methods of assessing components of the microbiome: metagenomics and metatranscriptomics, which interpret the microbial gene and transcriptase content, and metabolomics, which typifies the relationship and interactions between the microbiome and the host [1]. Medications and dietary changes can also alter the microbiome. Currently, however, it is unclear whether this variation in the microbiome is dysbiosis, which is a disease-related change in the microbiome, or whether this change is only a benign variability in the composition of the gut microbiome that is elicited by the underlying disease or medication. Dysbiosis has historically had a negative connotation, implying that disease states or medications can undesirably affect the microbiome [1]. On the contrary, variability alone would imply that the microbiome is dissimilar, but that this variability is not necessarily an adverse consequence. It is known that changes to the microbiome can occur through three specific mechanisms: outgrowth of pathobionts, loss of commensals, and loss of diversity. Outgrowth of pathobionts implies the overgrowth of potentially pathologic microbiota, whereas loss of commensals indicates a loss or reduction in normally residing microbiota. Loss of diversity occurs when there is diminished microbiota diversity in a given sample [1]. The aforementioned can occur with various disease states, secondary to environmental factors such as household and local environment, or in host-intrinsic factors including age and body mass index (BMI) and host-extrinsic factors such as

Esophageal Disease and the Role of the Microbiome
https://doi.org/10.1016/B978-0-323-95070-1.00011-4

medication, diet, and physical activity. Methods for assessing the microbiome can include amplicon-based sequencing (such as 16S rRNA gene sequencing, which includes targeting a genomic marker and then amplifying via polymerase chain reaction), as well as gathering and shearing microbiome DNA into minute fragments which are then sequenced and termed shotgun metagenomic sequencing. In addition, microbiome research has included animal studies that involve introducing a medication or organism into the animal and measuring the related functional effects. Previous studies of the microbiome have not clearly identified whether dysbiosis is the direct result of a disease state, medication, or diet, or whether there is simply an association with these things. Proton pump inhibitors (PPIs), which work by reducing gastric acid production and inhibiting hydrogen/potassium adenosine triphosphate pumps (ATP) on the parietal cells of the gastric lining, have recently been scrutinized for their role in dysbiosis and subsequent consequences such as *Clostridium difficile* infection. An important area of study, given the high prevalence of PPI use globally, is the potential effects that PPIs may have on the microbiome, with clarification as to whether any changes seen are correlational or causative. This chapter will focus to identify whether PPIs modulate the microbiome and what implications that would have for clinical practice. The primary areas of interest for animal and human studies of the microbiota of the gastrointenstinal (GI) tract are shown in Fig. 1.

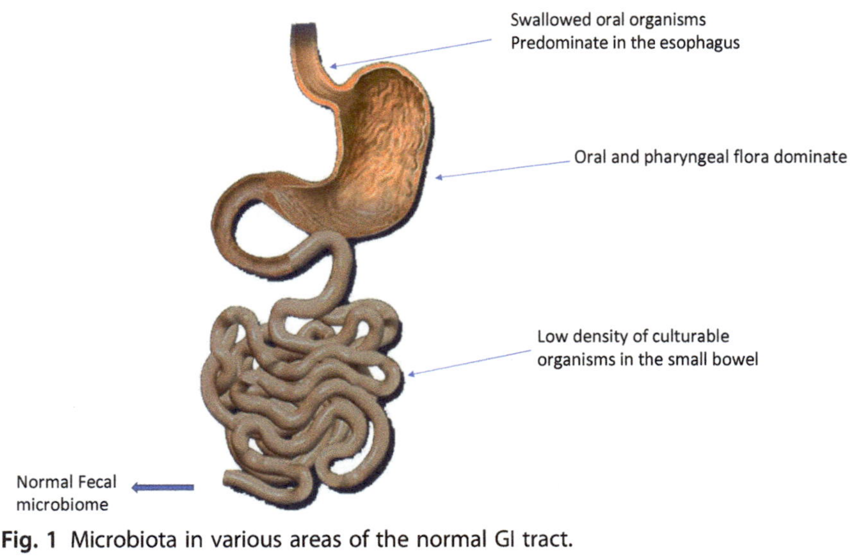

Fig. 1 Microbiota in various areas of the normal GI tract.

Animal studies

Both observational and experimental studies have been conducted on animals to determine the relationship between PPIs and the microbiome. An experimental study by Garcia-Mazcorro et al. involved the administration of omeprazole in eight healthy 9-month-old dogs [2]. The findings indicated that omeprazole did lead to quantitative changes in the gastrointestinal flora of healthy dogs, specifically with decrease in gastric *Helicobacter* species and higher *Firmicutes*, *Streptococcus*, and *Fusobacteria*, and an increase in total bacteria in the duodenum. There was an increase in the fecal *Lactobacillus* and *Faecalibacterium* species. The decrease in *Faecalibacterium* species occurred only in the male dogs and prompts interest given its known antiinflammatory properties in humans and its known depletion in patients with colitis. All of the dogs remained clinically healthy during the study, leading to the question of whether these microbiota variations are indeed harmful.

Experiments involving rats have also been conducted. A study involved the administration of lansoprazole to 24-week-old rats ($n = 11$) for 50 weeks with subsequent sequencing of the microbiota of the terminal ileum [3]. The findings showed no difference in stool frequency or consistency but an increase in *Firmicutes* of the *Clostridium g4* and *Lactobacillus* genera. The rats that were fed the lansoprazole-containing diet weighed less than the controls, indicating that long-term PPI use in rats may cause weight loss and alter the microbiota of the terminal ileum. This is a contradiction to human studies, which have associated a relative abundance of *Firmicutes* ssp. to *Bacteroidetes* ssp. with obesity and weight gain. A separate experimental study with rats involved the administration of omeprazole or lansoprazole in conjunction with nonsteroidal antiinflammatory drugs (NSAIDs) and determination of NSAID-induced intestinal injury in this population [4]. The authors determined that omeprazole-treated rats had a decrease in *Bifidobacteria* ssp. associated with development of enteric injury, but that the loss of commensal bacteria could be reduced with supplementation of *Bifidobacteria* ssp.–enriched suspension. The findings suggest that dysbiosis from PPIs may contribute to increased likelihood of NSAID-induced small bowel injury. However, the use of PPI's does offer protection against gastric ulcer formation in the stomach and proximal small intestine. Moreover, the use of PPIs alone did not have any effect on the intestinal tissue in spite of variability of the microbiota, and damage was only seen in patients on concurrent NSAIDs and PPIs. Gommers et al. investigated the diversity of the microbiome in rats on omeprazole and the association with the microbiota composition and hypomagnesemia [5]. Treatment with omeprazole did not damage the colonic mucosa, but there were increases in *Lactobacillus* ssp. and *Bifidobacterium* ssp. in the omeprazole treatment group. The experimental group also had lower magnesium levels, suggesting that there may be an association with PPI-induced microbiota alterations and absorption of ions, including calcium and magnesium, which was in some instances corrected with administration of prebiotic inulin fibers. Finally, an observational study by Kanno et al. investigating the association between rats and humans with hypochlorhydria and lower intestinal microflora suggested that reduced gastric acid leads to dysregulation of lower intestinal microflora [6]. This study did not specifically investigate the effects of PPIs on the microbiota but rather the association between reduced gastric acid and the microbiota, which would be a natural consequence of PPI therapy, given the acid-reducing properties.

Human studies

The microbiome of the GI tract varies in its composition depending on the organ being studied and the effects of both the disease state and acid inhibition on the microbial composition of the organ. The upper GI tract is influenced by ingested oral flora and gastric acid normally destroys many of the ingested organisms. Changes in gastric acid secretion can alter the environment of the stomach and allow organisms to pass into the small and large intestine affecting the microbiome (Fig. 2).

Salivary microbiome

Kawar et al. studied the salivary microbiome in patients with gastroesophageal reflux disease (GERD) treated with PPIs and found no significant differences in the salivary microbiome compared with healthy controls [7].

Esophageal microbiome

The microbiome of the esophagus is dominated by oral organisms that are swallowed. The use of PPIs has been postulated to affect the esophageal microbiome by altering the pH in the distal esophagus and by directly targeting P-type ATPase on bacterial proton pumps. Human studies using biopsy samples from healthy subjects and patients with esophagitis and Barrett's esophagus showed significant differences. Esophageal microbiomes can be classified into two types. The type I microbiome was dominated by the

Fig. 2 Changes in the microbiota of various areas of the normal GI tract caused by PPIs.

genus *Streptococcus* and concentrated in the phenotypically normal esophagus. Conversely, the type II microbiome contained a greater proportion of gram-negative anaerobes/microaerophiles and primarily correlated with esophagitis (odds ratio, 15.4) and Barrett's esophagus (odds ratio, 16.5) [8]. In a study of patients with chronic reflux disease or Barrett's esophagus treated with PPIs, Tasnim et al. found no significant association between the dose and duration of PPI use and the absolute levels of any organism and no significant changes in the esophageal microbial community [9]. Park et al. studied the effects of short-term PPI therapy in patients with nonerosive reflux disease [10]. The bacterial composition of the oropharynx and the esophagus was not affected by PPI use for 8 weeks.

Gastric microbiome

The gastric microbiome has been studied in individuals treated with PPIs with *Helicobacter pylori* infection and with disorders causing hypochlorhydria such as autoimmune gastritis. Samples from normal stomachs and patients treated with PPIs demonstrated similarly high microbial diversity. Patients with autoimmune atrophic gastritis also exhibited relatively high microbial diversity, but with samples dominated by *Streptococcus*, an organism associated with oral flora. Patients treated with PPI showed relatively few alterations in the gastric microbiota compared with healthy subjects [11]. A small study of dyspeptic patients found that patients taking PPIs had changes in the gastric microbiome with an increase in *Streptococcus* ssp. compared with untreated patients [12]. In a study of 45 patients taking PPIs, 16s RNA gene profiling was used to study the bacterial communities in stimulated saliva, gastric fluid, and feces [13]. Shi et al. studied the gastric microbial composition in GERD patients taking PPIs and found a reduction in bacterial diversity and a higher abundance of *Streptococcaceae* ssp. in PPI users [14]. In cirrhotic patients in whom PPI therapy was initiated, a significant increase in the abundance of *Porphyromonadaceae* ssp. and *Streptococcaceae* ssp. was seen [15].

Small and large bowel

The duodenum is an area of interest for patients with dyspeptic symptoms. A study evaluating PPI use in dyspeptic patients found that short-term PPI therapy had no significant effect in patients with functional dyspepsia but *Streptococcus* ssp. became more abundant in healthy controls after PPI therapy [16].

Studies on the small bowel have focused primarily on bacterial overgrowth in the small intestine diagnosed indirectly by breath testing. The results of these studies have been conflicting and are confounded by small sample sizes, heterogenous populations, and the inherent limitations of glucose breath testing. Studies of duodenal aspirates provide a more reliable method of assessing alterations to the microbiome.

Studies using breath testing have yielded conflicting results. A retrospective study of more than 1000 patients found no association between PPI use and small bowel bacterial overgrowth [17]. A smaller study of 450 patients and controls found that the prevalence of small bowel bacterial overgrowth defined by glucose breath testing was 50% and was significantly higher than in non-PPI users [18].

In a recent study using 16s DNA testing in duodenal aspirates and fecal samples, Weitsman et al. found that the duodenal microbiome exhibited no phylum-level differences between PPI users and non-PPI users, but demonstrated significantly higher relative abundances of families *Campylobacteraceae* ssp. and *Bifidobacteriaceae* ssp., and lower relative abundance of *Clostridiaceae* ssp., in PPI subjects [19]. Small bowel bacterial overgrowth rates were not significantly different between groups, whether defined by culture ($>10^3$ CFU/mL) or 16S sequencing and were not different between subjects taking different PPIs. The stool microbiome exhibited significantly higher abundance of family *Streptococcaceae* and a reduction in *Clostridiaceae*, the significance of which is unknown.

A recent meta-analysis of studies found a modestly increased risk for small bowel bacterial overgrowth in patients using PPIs (OR 1.71, 95% confidence interval 1.20–2.43) but the confidence intervals in the range where residual confounding could account for the results [20].

Fecal microbiome studies

Studies on fecal samples have been performed to determine the effect of PPIs on the fecal microbiome. A study of three independent Dutch cohorts found a decrease in species richness and diversity and statistical differences in many bacterial taxa with overrepresentation of oral organisms such as *Streptococcaceae* [21]. In a large twin cohort study in the United Kingdom, the association between PPI use and changes in the microbiome was investigated in 1827 healthy twins. Similar to the Dutch study, the authors found a lower abundance of gut commensals and lower microbial diversity in PPI users and an overabundance of oral and upper gastrointestinal commensals particularly *Streptococcaceae* ssp. [22].

Other observational studies have yielded similar results [23–25]. In healthy volunteers treated with PPIs, the ratio of *Streptococcus* organisms increased in fecal samples consistent with studies in patients with GERD [26].

Small interventional studies in adults and children suggest that short-term use of PPIs does not cause permanent alterations in the fecal microbiome and that discontinuation of the PPI results in a return of the microbiome to its original state [27–29].

A systematic review summarized the data until 2019 and concluded that in the upper GI tract PPI users showed a significant overgrowth of orally derived bacteria but the effects in fecal samples were more complicated [30]. There were increases in multiple taxa from the family *Bacillales* and reductions in others such as *Bifidobacteriaceae*. There was significant heterogeneity in the studies and the definitions of PPI users was highly variable. Some studies evaluated current or recent users of PPIs, whereas others allowed entry to patients who had used PPIs in the preceding 3 years without specifying a duration.

Potassium competitive acid inhibitors are a new class of acid inhibitor that cause more profound acid suppression than PPIs. Limited data are available on the effect of these agents on the microbiome. One small study found long-term changes in the microbiome after *H. pylori* eradication therapy with this agent, but it remains unclear whether the effects were those of acid suppression or the antimicrobials used in the treatment regimen [31].

Limitations of currently available data

Currently available data are primarily observational and are limited by several factors. The sample size of the studies is small, and in most studies, richness and diversity were not significantly altered but changes in various taxa were noted. The increase in these taxae does not represent an increase in pathobionts but rather a change in commensals, and the significance of this change remains unclear and associations with disease states remain speculative.

Disease associations

Alterations in the microbiome of the GI tract have been postulated to cause disease in various parts of the GI tract but these remain speculative. Fig. 3 summarizes the associations and possible effects of PPIs. For example, the effects of high-dose PPI therapy in preventing dysplasia and cancer in Barrett's esophagus may be related to alterations in the esophageal microbiome [32]. The persistence of a risk for gastric cancer despite *H. pylori* eradication may be related to alterations in the gastric microbiome and PPIs could

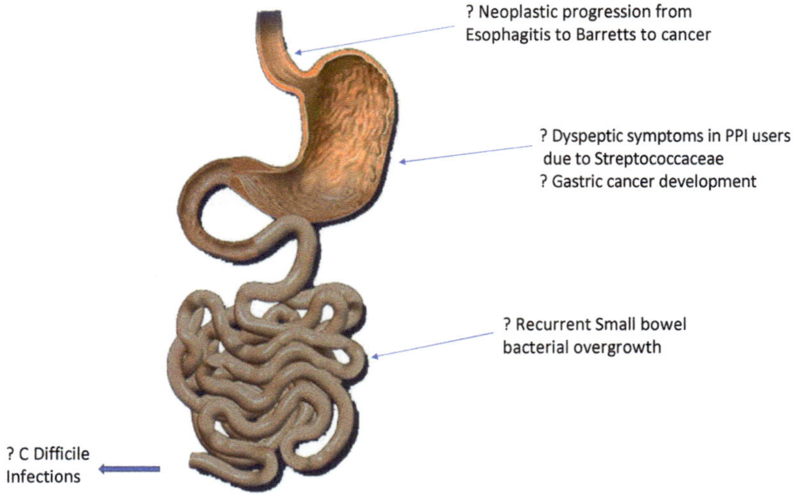

Fig. 3 Postulated disease state associations with alterations in the microbiota caused by PPIs.

modify this risk [33]. Recurrent small bowel bacterial overgrowth occurs in patients on and off PPIs and the role of acid inhibition remains unclear. Finally, *C. difficile* colitis has been clearly linked to PPI use, but it remains unclear whether this is due to alterations in the microbiome or to the passage of spores through the stomach where they would ordinarily have been destroyed by gastric acid [34].

Summary

PPIs have an effect on the microbiota of the upper GI and lower GI tract. In the upper GI tract, the effect is to change the preponderance of the organisms toward the organisms of the oropharynx with an increase in *Streptococcaceae* ssp. Some studies show a decrease in the diversity of the gastric microbiota, whereas the lower GI tract shows inconsistent changes in taxa. The significance of these changes and their relationship to disease remains speculative.

Future priorities

1. *Correct weakness in study design.* Currently published studies are small in size consisting of less than 20 patients. Large population-based studies are opportunistic and therefore have entry and exclusion criteria that are not clinically relevant.
2. *Study the effects of PPIs on bacterial metabolic pathways and products of these pathways.*
3. *Correlate changes in bacterial metabolic pathways and products of these pathways with disease pathogenesis.*
4. *Evaluate if acid-related microbial changes are related to thresholds of acid suppression and correlate related responses (e.g., PPIs versus H2 receptor antagonists, surgical- or medical-related conditions of achlorhydria).*

References

[1] M. Camilleri, N. Vakil, Microbiome: in search of mechanistic information and relevance, Am. J. Gastroenterol. 114 (7) (2019) 1014–1016, https://doi.org/10.14309/ajg.0000000000000306. 31211707.

[2] J.F. Garcia-Mazcorro, J.S. Suchodolski, K.R. Jones, et al., Effect of the proton pump inhibitor omeprazole on the gastrointestinal bacterial microbiota of healthy dogs, FEMS Microbiol. Ecol. 80 (3) (2012) 624–636, https://doi.org/10.1111/j.1574-6941.2012.01331.x. Epub 2012 Mar 12 22324305.

[3] C.M. Shin, N. Kim, Y.S. Kim, et al., Impact of long-term proton pump inhibitor therapy on gut microbiota in F344 rats: pilot study, Gut Liver 10 (6) (2016) 896–901, https://doi.org/10.5009/gnl15529. 27458177.

[4] J.L. Wallace, S. Syer, E. Denou, et al., Proton pump inhibitors exacerbate NSAID-induced small intestinal injury by inducing dysbiosis, Gastroenterology 141 (4) (2011) 1314–1322. 1322.e1–5 https://doi.org/10.1053/j.gastro.2011.06.075. Epub 2011 Jul 13 21745447.

[5] L.M.M. Gommers, T.H.A. Ederveen, J. van der Wijst, et al., Low gut microbiota diversity and dietary magnesium intake are associated with the development of PPI-induced hypomagnesemia, FASEB J. 33 (10) (2019) 11235–11246, https://doi.org/10.1096/fj.201900839R. Epub 2019 Jul 12.

[6] T. Kanno, T. Matsuki, M. Oka, et al., Gastric acid reduction leads to an alteration in lower intestinal microflora, Biochem. Biophys. Res. Commun. 381 (4) (2009) 666–670, https://doi.org/10.1016/j.bbrc.2009.02.109. Epub 2009 Feb 25.

[7] N. Kawar, S.G. Park, J.L. Schwartz, et al., Salivary microbiome with gastroesophageal reflux disease and treatment, Sci. Rep. 11 (1) (2021) 188, https://doi.org/10.1038/s41598-020-80170-y.

[8] L. Yang, X. Lu, C.W. Nossa, F. Francois, R.M. Peek, Z. Pei, Inflammation and intestinal metaplasia of the distal esophagus are associated with alterations in the microbiome, Gastroenterology 137 (2) (2009) 588–597, https://doi.org/10.1053/j.gastro.2009.04.046. Epub 2009 Apr 23.

[9] S. Tasnim, A.L. Miller, D.C. Jupiter, et al., Effects of proton pump inhibitor use on the esophageal microbial community, BMC Gastroenterol. 20 (1) (2020) 312, https://doi.org/10.1186/s12876-020-01460-3.

[10] C.H. Park, S.I. Seo, J.S. Kim, et al., Treatment of non-erosive reflux disease and dynamics of the esophageal microbiome: a prospective multicenter study, Sci. Rep. 10 (1) (2020) 15154, https://doi.org/10.1038/s41598-020-72082-8. 32938975.

[11] B.N. Parsons, U.Z. Ijaz, R. D'Amore, et al., Comparison of the human gastric microbiota in hypochlorhydric states arising as a result of *Helicobacter pylori*-induced atrophic gastritis, autoimmune atrophic gastritis and proton pump inhibitor use, PLoS Pathog. 13 (11) (2017) e1006653, https://doi.org/10.1371/journal.ppat.1006653. eCollection 2017 Nov.

[12] F. Paroni Sterbini, A. Palladini, L. Masucci, et al., Effects of proton pump inhibitors on the gastric mucosa-associated microbiota in dyspeptic patients, Appl. Environ. Microbiol. 82 (22) (2016) 6633–6644, https://doi.org/10.1128/AEM.01437-16. Print 2016 Nov 15 27590821.

[13] A. Tsuda, W. Suda, H. Morita, et al., Influence of proton-pump inhibitors on the luminal microbiota in the gastrointestinal tract, Clin. Transl. Gastroenterol. 6 (6) (2015) e89, https://doi.org/10.1038/ctg.2015.20. 26065717.

[14] Y.C. Shi, S.T. Cai, Y.P. Tian, et al., Effects of proton pump inhibitors on the gastrointestinal microbiota in gastroesophageal reflux disease, Genom. Proteom. Bioinform. 17 (1) (2019) 52–63, https://doi.org/10.1016/j.gpb.2018.12.004. Epub 2019 Apr 25 31028880.

[15] J.S. Bajaj, C. Acharya, A. Fagan, et al., Proton pump inhibitor initiation and withdrawal affects gut microbiota and readmission risk in cirrhosis, Am. J. Gastroenterol. 113 (8) (2018) 1177–1186, https://doi.org/10.1038/s41395-018-0085-9. Epub 2018 Jun 6.

[16] L. Wauters, R.Y. Tito, M. Ceulemans, et al., Duodenal dysbiosis and relation to the efficacy of proton pump inhibitors in functional dyspepsia, Int. J. Mol. Sci. 22 (24) (2021) 13609, https://doi.org/10.3390/ijms222413609. 34948413.

[17] S.K. Ratuapli, T.G. Ellington, M.T. O'Neill, et al., Proton pump inhibitor therapy use does not predispose to small intestinal bacterial overgrowth, Am. J. Gastroenterol. 107 (5) (2012) 730–735, https://doi.org/10.1038/ajg.2012.4. Epub 2012 Feb 14 22334250.

[18] L. Lombardo, M. Foti, O. Ruggia, A. Chiecchio, Increased incidence of small intestinal bacterial overgrowth during proton pump inhibitor therapy, Clin. Gastroenterol. Hepatol. 8 (6) (2010) 504–508, https://doi.org/10.1016/j.cgh.2009.12.022. Epub 2010 Jan 6 20060064.
[19] S. Weitsman, S. Celly, G. Leite, et al., Effects of proton pump inhibitors on the small bowel and stool microbiomes, Dig. Dis. Sci. 67 (1) (2022) 224–232, https://doi.org/10.1007/s10620-021-06857-y. Epub 2021 Feb 3 33534012.
[20] T. Su, S. Lai, A. Lee, X. He, S. Chen, Meta-analysis: proton pump inhibitors moderately increase the risk of small intestinal bacterial overgrowth, J. Gastroenterol. 53 (1) (2018) 27–36, https://doi.org/10.1007/s00535-017-1371-9. Epub 2017 Aug 2.
[21] F. Imhann, M.J. Bonder, A. Vich Vila, et al., Proton pump inhibitors affect the gut microbiome, Gut 65 (5) (2016) 740–748, https://doi.org/10.1136/gutjnl-2015-310376. Epub 2015 Dec 9 26657899.
[22] M.A. Jackson, J.K. Goodrich, M.E. Maxan, et al., Proton pump inhibitors alter the composition of the gut microbiota, Gut 65 (5) (2016) 749–756, https://doi.org/10.1136/gutjnl-2015-310861. Epub 2015 Dec 30.
[23] A.G. Clooney, C.N. Bernstein, W.D. Leslie, et al., A comparison of the gut microbiome between long-term users and non-users of proton pump inhibitors, Aliment. Pharmacol. Ther. 43 (9) (2016) 974–984, https://doi.org/10.1111/apt.13568. Epub 2016 Feb 29.
[24] M. Hojo, T. Asahara, A. Nagahara, et al., Gut microbiota composition before and after use of proton pump inhibitors, Dig. Dis. Sci. 63 (11) (2018) 2940–2949, https://doi.org/10.1007/s10620-018-5122-4. Epub 2018 May 24.
[25] T. Takagi, Y. Naito, R. Inoue, et al., The influence of long-term use of proton pump inhibitors on the gut microbiota: an age-sex-matched case-control study, J. Clin. Biochem. Nutr. 62 (1) (2018) 100–105, https://doi.org/10.3164/jcbn.17-78. Epub 2017 Dec 12.
[26] T. Mishiro, K. Oka, Y. Kuroki, et al., Oral microbiome alterations of healthy volunteers with proton pump inhibitor, J. Gastroenterol. Hepatol. 33 (5) (2018) 1059–1066, https://doi.org/10.1111/jgh.14040. Epub 2018 Feb 14 29105152.
[27] C.T. Seto, P. Jeraldo, R. Orenstein, N. Chia, J.K. DiBaise, Prolonged use of a proton pump inhibitor reduces microbial diversity: implications for *Clostridium difficile* susceptibility, Microbiome 2 (2014) 42, https://doi.org/10.1186/2049-2618-2-42. eCollection 2014.
[28] C. Castellani, G. Singer, K. Kashofer, et al., The influence of proton pump inhibitors on the fecal microbiome of infants with gastroesophageal reflux—a prospective longitudinal interventional study, Front. Cell. Infect. Microbiol. 7 (2017) 444, https://doi.org/10.3389/fcimb.2017.00444. eCollection 2017.
[29] S.H. Koo, J. Deng, D.S.W. Ang, et al., Effects of proton pump inhibitor on the human gut microbiome profile in multi-ethnic groups in Singapore, Singap. Med. J. 60 (10) (2019) 512–521, https://doi.org/10.11622/smedj.2018152. Epub 2018 Nov 29.
[30] L. Macke, C. Schulz, L. Koletzko, P. Malfertheiner, Systematic review: the effects of proton pump inhibitors on the microbiome of the digestive tract-evidence from next-generation sequencing studies, Aliment. Pharmacol. Ther. 51 (5) (2020) 505–526, https://doi.org/10.1111/apt.15604. Epub 2020 Jan 28 31990420.
[31] S. Suzuki, T. Gotoda, C. Takano, et al., Long term impact of vonoprazan-based *Helicobacter pylori* treatment on gut microbiota and its relation to post-treatment body weight changes, Helicobacter 26 (6) (2021) e12851, https://doi.org/10.1111/hel.12851. Epub 2021 Sep 6 34486195.
[32] R.S. Bresalier, Chemoprevention of Barrett's esophagus and esophageal adenocarcinoma, Dig. Dis. Sci. 63 (8) (2018) 2155–2162, https://doi.org/10.1007/s10620-018-5149-6. 29948566.

[33] K.S. Cheung, E.W. Chan, A.Y.S. Wong, L. Chen, I.C.K. Wong, W.K. Leung, Long-term proton pump inhibitors and risk of gastric cancer development after treatment for *Helicobacter pylori*: a population-based study, Gut 67 (1) (2018) 28–35, https://doi.org/10.1136/gutjnl-2017-314605. Epub 2017 Oct 31 29089382.
[34] G.I. Leontiadis, M.A. Miller, C.W. Howden, How much do PPIs contribute to *C. difficile* infections? Am. J. Gastroenterol. 107 (7) (2012) 1020–1021, https://doi.org/10.1038/ajg.2012.174. 22764024.

CHAPTER 4

Esophageal dysbiosis and pathogenesis of gastroesophageal reflux disease

Steve M. D'Souza[a], Manasa Vallabhaneni[a], Kevin V. Houston[b], Lindsey Cundra[c], Ana Vilela[a], Byung Soo Yoo[d], and David A. Johnson[e]
[a]Department of Internal Medicine, Eastern Virginia Medical School, Norfolk, VA, United States
[b]Department of Internal Medicine, Virginia Commonwealth University, Richmond, VA, United States
[c]Department of Internal Medicine, Georgetown University, Washington, DC, United States
[d]Division of Gastroenterology, Department of Internal Medicine, Carolinas Medical Center, Charlotte, NC, United States
[e]Division of Gastroenterology, Department of Internal Medicine, Eastern Virginia Medical School, Norfolk, VA, United States

Introduction

Gastroesophageal reflux disease (GERD) is a condition of symptoms or pathologic changes associated with the retrograde flow of gastric contents into the esophagus [1]. While some extent of gastroesophageal reflux may be normal, symptoms caused by increased acid exposure and/or the presence of mucosal injury point toward GERD [1]. With the advent of newer tools to characterize the gut microbiome, understanding of factors that contribute to esophageal disease has expanded, and it has become clear that the local flora plays a significant role. In this chapter, we discuss the evolving changes in the understanding of the pathogenesis of GERD.

Traditional view of GERD progression

Initial belief regarding pathogenesis suggested that acid damage to esophageal squamous epithelium leads to increased intercellular space and esophageal permeability [2]. This damage was perceived to manifest as heartburn and mucosal injury leading to histological changes including chronic inflammation and possibly to metaplasia.

https://doi.org/10.1016/B978-0-323-95070-1.00003-5

Anatomic factors

Although the exact pathogenesis of GERD remains unclear. Well-recognized, important risk factors include anatomical variations, such as lower esophageal sphincter (LES) incompetence, hiatal hernias, and duodeno-gastroesophageal reflux [3]. The LES incompetence or hypotensive lower esophageal sphincter is defined as a lower esophageal sphincter pressure <10 mmHg [4]. This incompetence is associated with increased reflux due to easy passage of gastric contents during times of low sphincter pressure, in particular when associated with a hiatal hernia. Transient lower esophageal relaxation (TLESR) is esophageal relaxation despite lack of swallow stimulus. This is the most common pathogenic mechanism associated with GERD, especially in the absence of a hiatal hernia [1].

Although TLESR does not differ among healthy and GERD patients, TLESR is more likely associated with acid reflux in GERD patients [5]. Hiatal hernias can lead to decreased lower esophageal sphincter pressure and increased GERD frequency. The degree of relaxation of the lower esophageal sphincter is directly proportional to the distance between the squamocolumnar junction and hiatal hernia [6]. The hernia moves the LES above the diaphragm, removing the diaphragmatic contribution to the LES pressure, as well as leading to a compressive "pinchcock" effect where the acid cannot easily reenter the gastric space [7]. Other factors leading to variations in gastroesophageal junction anatomy, such as obesity or abrupt coughing, can modify protective mechanisms against reflux by increasing intraabdominal pressure and allowing passive reflux into the esophagus. These pathologic changes lead to mucosal injury that cannot be overcome by physiologic defense mechanisms, including acid clearance and mucosal integrity.

Nonanatomic factors

Nonanatomic factors can also contribute to GERD progression. One factor is xerostomia or decreased saliva production. Hyposalivation can lead to GERD due to decreased acid neutralization in the esophagus by saliva, as well as decreased swallowing effect of primary peristalsis and effective clearance of gastric refluxate [8].

The degree of damage to the mucosa is proportional to the frequency and duration of reflux events along with the patient's phenotypic predisposition to the mucosal response [9]. Additionally, the components of the gastric refluxate may contribute to the degree of damage. Notably, the pH, presence of pepsin, and/or presence of bile acids are recognized as important pathogenic factors [10,11]. Gastric acid with a pH <2 and pepsin has been associated with the most severe esophageal damage by binding to laryngeal cells, entering the cell through endocytosis, and bringing down cellular defense mechanisms [11,12]. Bile acids work synergistically with acid and trypsin in conjugated and unconjugated forms, respectively. Both forms of bile acid can cause severe damage through signaling mechanisms, such as vascular endothelial growth factor (VEGF), even leading to metaplastic and neoplastic transformation [13].

The caustic effects of reflux progress from epithelial cell damage to esophagitis through various mechanisms. Esophagitis is an inflammatory process marked by the presence of lymphocytic infiltrate and increased intracellular spaces [2]. Earlier theories of reflux esophagitis suggested that epithelial cell damage from GERD propagates an inflammatory response through direct damage [14]. Both animal and human studies, however, have suggested that progression to esophagitis occurs due to gastric acid stimulating esophageal epithelial cells to secrete inflammatory cytokines, such as interleukin-8 and interleukin-1β [15,16]. Various studies have shown that genetic polymorphisms in interleukin-1β genes can predispose patients with GERD to become more likely to develop reflux esophagitis due to hyperacidity [17,18]. Esophagitis can eventually lead to complications such as Barrett's esophagus, stricture, ulceration, and possible hemorrhage [19].

Revised mechanism for GERD pathogenesis

There are several factors in the traditional understanding of GERD pathogenesis remain unclear. While reflux of gastric acid and bile is related to esophagitis, direct caustic effects on epithelium may not be the main driving factor. This can be demonstrated by biopsies taken of reflux esophagitis demonstrating intact mucosa with submucosal inflammation [15]. If caustic exposure was the primary driving factor, mucosal damage would be necessary to cause submucosal effects. Additionally, caustic exposure would not explain why NERD with significant acid exposure commonly does not progress to reflux esophagitis [20].

One proposed mechanism is that luminal factors act as a trigger for submucosal changes that initiate pathogenesis within a suitable environment. Acid and bile exposure may cause activation of local nerves, leading to pain, and inflammatory marker secretion [21]. This, in combination with a conducive local climate caused by prior inflammation or proinflammatory microbiome-mucosal interactions, leads to tissue injury and disease progression. These luminal triggers cause lymphocytic invasion into the submucosa and ultimately an inflammatory cascade toward the mucosa. Essentially, what is typically seen is an outside-in approach to pathogenesis rather than inside out.

Microbiome changes

A major factor that affects the progression of reflux to histological disease is the local environment. The luminal milieu produced by the composition of local microbes, as well as associated gut-microbial interactions may affect this risk. Several microbiome changes have been identified in to be related to GERD [22,23].

Overall composition

Initial work studying the "normal" esophageal microbiome found the most prevalent genera to be *Streptococcus*, *Prevotella*, and *Veillonella* [24,25]. Significant interindividual variation exists because of the number of factors that affect composition (see Chapter 12). In 2009, investigators evaluating compositional differences between normal and disease states, found two compositional phenotypes (normal and pathological), referred to as Type I and Type II [26]. Type I microbiomes were composed of primarily gram-positive flora, with a high incidence of Firmicutes, most commonly *Streptococcus* spp., and were considered to be in the nomal state. Type II microbiomes were associated with an increased risk of GERD and BE, and had similar taxa but differed in compositional variation. This was most manifest as a higher abundance of gram-negative organisms and anaerobes, most notably *Prevotella, Veillonella, Neisseria*, and *Haemophilus* spp. *Streptococcus* spp. remained the most abundant taxa, however, was found to have decreased relative abundance in pathological states [27].

It is important to note that there are two key findings in the microbiomes within disease states. First, the type of flora did not vary significantly. Within interindividual variance, similar flora was consistently present [28]. Rather, it was the relative abundance of this flora that was associated with esophageal disease. Second, disease-associated microbiome changes precede histological changes, suggesting that they may play a role in pathogenesis [29].

Microgeographic changes

Variation in the local microbiome is also present as a gradient from the proximal to distal esophagus. The proximal esophageal flora is most similar to

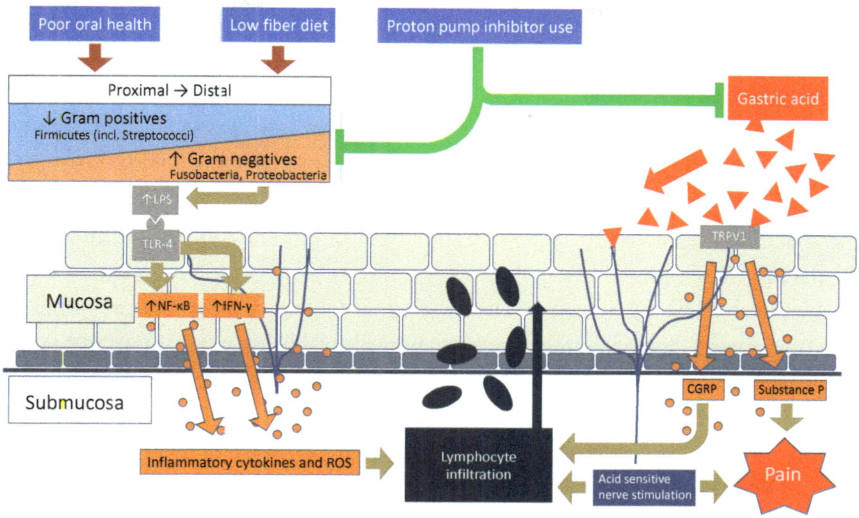

Fig. 1 Environmental factors alter the local esophageal microbiome, which normally has a gram-positive to gram-negative gradient, toward increased proportion of gram-negatives. Activation of TLR-4 by gram-negative lipopolysaccharide leads to an inflammatory cascade that results in lymphocyte infiltration. In addition, gastric acid activates TRPV1 on local nerve fibers and results in CGRP and substance P expression, which contributes to the local inflammatory response as well as pain. *CGRP*, calcitonin gene-related peptide; *IFN*, interferon; *LPS*, lipopolysaccharide; *NF-κB*, nuclear factor-kappa beta; *ROS*, reactive oxygen species; *TLR*, toll-like-receptor; *TRPV1*, transient receptor potential cation channel subfamily V member 1. *(Reproduced with permission from S.M. D'Souza, K. Houston, L. Keenan, B.S. Yoo, P.J. Parekh, D.A. Johnson, Role of microbial dysbiosis in the pathogenesis of esophageal mucosal disease: a paradigm shift from acid to bacteria?, World J. Gastroenterol. 27 (18) (2021) 2054–2072. https://doi.org/10.3748/wjg.v27.i18.2054.)*

oropharyngeal flora, with the greatest diversity present, and a decrease in diversity aborally [30]. As the luminal environment becomes progressively harsher distally, with exposure to significant chemical factors—including acid, bile salts, and proteases, the local flora needs more intrinsic protection [31–36]. Thus, the local flora is more likely to have lipopolysaccharide (LPS) (Gram negative) and be spore-forming, and includes taxa such as *Prevotella*, with a marked decrease in *Streptococcus* abundance from proximal to distal (Fig. 1) [37–40].

Biofilm

Biofilm is a collection of microbes with the synergistic expression of gene products to provide maximal protection and adherence [41]. These colonies allow for a proliferation of certain flora within harsher environments and produce a protective, organized extracellular matrix [42]. Biofilm has been proposed as a mechanism through which some bacterial taxa maintain a continuous presence along mucosal surfaces [42,43]. This phenomenon has been demonstrated elsewhere in the GI tract, mainly the colon, and has been associated with several conditions [44–47]. Recent literature is demonstrating its role in providing stability to the microbial composition in the harsher climate of the lower esophagus and the effect of altered biofilm composition on pathogenesis [48].

Just as there is a gradient of compositional variation from the proximal to distal esophagus, the presence ofbiofilm may also indicate variation in the composition offlora from the central lumen to the mucosal surface, especially within the distal esophagus, as there is less clearance of microbes that are contained within the biofilm. Some evidence exists that the presence of certain flora within esophageal biofilm is associated with GERD and BE [48]. One investigation found an increased presence of *Campylobacter* spp., mainly *C. concisus* [48]. More work is needed to investigate compositional differences in luminal versus mucosal-associated flora and the impact on esophageal disease.

Intestinal microbiome

The intestinal microbiome has been implicated as playing a role in the development of many conditions, including several disorders of gut-brain interaction, inflammatory conditions, and malignancies [49]. This occurs through the modulatory effect of various bacterial products on many neurological and inflammatory functions. In addition to the implications of local dysbiosis within the esophagus, intestinal dysbiosis may also be related to GERD pathogenesis.

Several investigations have found that colonic microbial compositional changes may be related to stress responsiveness, nociception, and food intake [50,51]. Changes in the colonic microbiome have been associated with the presence of obesity and systemic inflammation, especially when correlated to the *Firmicutes/Bacteroidetes* ratio [52]. This occurs through a similar pathway to local esophageal pathogenesis. Increased production of endotoxin from gram-negative flora and its binding to surface TLR-4 triggers a downstream cascade of local and systemic inflammatory activation [53]. This has effects on the development of inflammation elsewhere in the body, behavior and visceral pain perception, and overall GI function, including motility [50]. Murine models have demonstrated that hypernociception is reversibly decreased in germ-free mice as a result of decreased gut-immune crosstalk [54].

Other bacterial products may also be related to GERD pathogenesis (Fig. 2). Fermentation of oligo-, di-, monosaccharides, and polyols

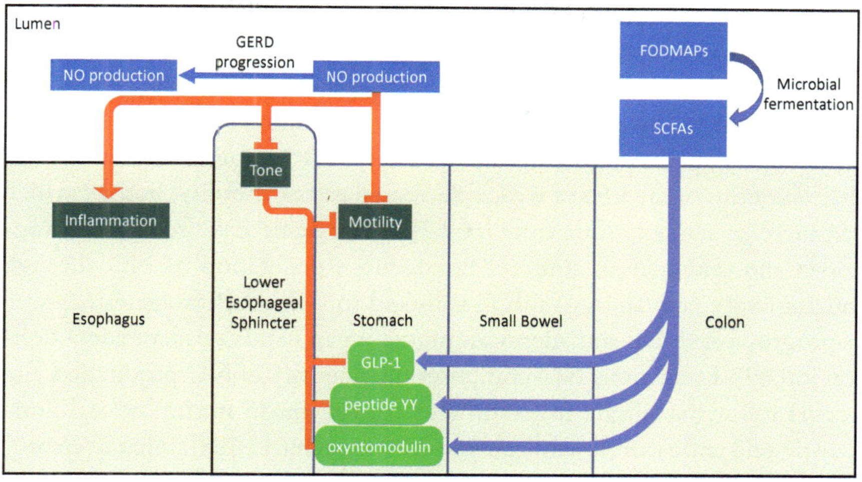

Fig. 2 GERD progression leads to more proximal production of luminal nitric oxide. This leads to esophageal inflammation, decreased LES tone, and gastric motility. Fermentation of FODMAPs to SCFAs in the colon leads to GLP-1, peptide YY, and oxyntomodulin expression, which inhibit LES tone and gastric motility. *FODMAPs*, fermentable oligo-, di-, monosaccharides, and polyols; *GERD*, gastroesophageal reflux disease; *GLP-1*, glucagon-like peptide 1; *LES*, lower esophageal sphincter; *NO*, nitric oxide; *SCFAs*, short chain fatty acids.

(FODMAPs) results in the production of short-chain fatty acids (SCFA) [55,56]. This colonic fermentation is associated with increased L-cell production of GLP-1 and has been demonstrated to increase the frequency of TESLR, acid-exposure time, as well as symptoms of GERD [57,58]. Both oral intake of nondigestible oligosaccharides, as well as intracolonic infusion of oligosaccharides or SCFAs, have been associated with decreased gastric tone through the increase in peptide YY, which regulates gastric and intestinal motility [59,60]. This mechanism has been referred to as an "ileocolonic brake," in that distal SCFA production has effects on proximal motility [61]. Colonic fermentation of oligosaccharides and SCFA presence have also been associated with a decrease in postprandial LES pressure lower than baseline, through increased oxyntomodulin and peptide YY production [58,62].

Inflammatory pathway

One possible mechanism for dysbiosis-related contribution to pathogenesis may be the LPS-TLR-4 pathway (Fig. 1). Increased LPS in the distal esophagus through changes outlined earlier, stimulates a submucosal inflammatory cascade by binding to epithelial TLR-4, causing NF-κB and subsequently inflammatory cytokine expression interleukin (IL)-1B, IL-8, IL-18, and tumor necrosis factor alpha, leading to lymphocytic proliferation and localized inflammatory response [63,64].

Additionally, cytokine expression has been linked to inducible nitric oxide (NO) synthase (iNOS) activity [65]. NO has been associated with the relaxation of the LES as well as decreased gastric motility, both of which can increase acid exposure time [66,67]. It can be derived from both exogenous and endogenous sources. Endogenous generation of NO through inflammatory activation of iNOS is linked to GERD pathogenesis as well as progression to BE and adenocarcinoma, in an exposure-dependent fashion [68,69]. Exogenous NO comprises the majority of NO production and occurs following nitrate ingestion and conversion to nitrite by oral flora. Gastric acid and ascorbic acid generate NO from nitrite [68]. It has been proposed that reflux of gastric contents may lead to increased luminal NO production in the distal esophagus, where it may contribute to local sphincter relaxation, decreased musical barrier function and permeability, and subsequent inflammatory activation (Fig. 2) [68,70].

In addition to microbe-gut interactions, a separate intertwined system induces an esophageal inflammatory response to acid and bile exposure. Gastric refluxate causes activation of acid-sensitive ion channels (ASIC), transient receptor potential cation channel subfamily V member 1 (TRPV1), and protease-activated receptor 2 (PAR2) [71]. Patients with reflux symptoms, including those without esophagitis (NERD) have increased expression of acid-sensitive receptors, and it is believed that this may be a component of visceral hypersensitivity. ASIC activation is associated with pain sensation and increased epithelial permeability [72,73]. TRPV1, the capsaicin receptor, expression on local C fibers has been linked to decreased barrier function [74,75]. PAR2, in combination with TRPV1, costimulates the subsequent inflammatory response through the release of calcitonin gene-related peptide and substance P, which contribute to neurokinin 1 receptor and subsequent downstream activation of the inflammatory cascade as well as the local sensation of pain [21,76].

Conclusions

The pathogenesis of GERD is complex and multifactorial. The traditional heretofore understanding of the pathogenesis of GERD, has been inflammation caused by the surface mucosal caustic effects of acid. Recent work suggests that GERD is a result of the direct and indirect effects of acid and bile reflux, as well as a shift in the esophageal biome which may be a previously unrecognized key factor. Clarification of an infectious component to the pathogenesis of GERD may have profound effects on both testing and disease mitigation strategies.

Future directions

- It remains unclear if esophageal microbiomic changes precede inflammatory changes seen on histology, and/or may predict and contribute to the progression of disease.
- The role of acid and other gastric refluxate with specific effects on the esophageal microbiome needs clarification.
- Specific cytokine and immunologic responses associated with esophageal dysbiosis biomic mitigation have to be clarified.
- If microbial dysbiosis is identified, mitigation strategies have to be developed as well as symptomatic and physiologic assessments for these therapeutic interventions.
- Safety and durability of effect for any dysbiosis-directed therapies have to be assessed. Other potential confounders for esophageal dysbiosis need clarification—e.g., role of diet, obesity, smoking, sleep, and disease states.

Author contributions

Johnson DA contributed to the construction of the project. All authors wrote and edited the manuscript.

Conflicts of interest

None.

References

[1] J. Tack, J.E. Pandolfino, Pathophysiology of gastroesophageal reflux disease, Gastroenterology 154 (2) (2018) 277–288, https://doi.org/10.1053/j.gastro.2017.09.047.
[2] N.A. Tobey, J.L. Carson, R.A. Alkiek, R.C. Orlando, Dilated intercellular spaces: a morphological feature of acid reflux—damaged human esophageal epithelium, Gastroenterology 111 (5) (1996) 1200–1205, https://doi.org/10.1053/gast.1996.v111.pm8898633.
[3] K.H. Fuchs, A.M. Lee, W. Breithaupt, G. Varga, B. Babic, S. Horgan, Pathophysiology of gastroesophageal reflux disease—which factors are important? Transl. Gastroenterol. Hepatol. 6 (2021) 53, https://doi.org/10.21037/tgh.2020.02.12.

[4] P.J. Kahrilas, W.J. Dodds, W.J. Hogan, M. Kern, R.C. Arndorfer, A. Reece, Esophageal peristaltic dysfunction in peptic esophagitis, Gastroenterology 91 (4) (1986) 897–904, https://doi.org/10.1016/0016-5085(86)90692-x.
[5] P.J. Kahrilas, G. Shi, M. Manka, R.J. Joehl, Increased frequency of transient lower esophageal sphincter relaxation induced by gastric distention in reflux patients with hiatal hernia, Gastroenterology 118 (4) (2000) 688–695, https://doi.org/10.1016/s0016-5085(00)70138-7.
[6] T. Hershcovici, H. Mashimo, R. Fass, The lower esophageal sphincter, Neurogastroenterol. Motil. 23 (9) (2011) 819–830, https://doi.org/10.1111/j.1365-2982.2011.01738.x.
[7] J. Richter, Do we know the cause of reflux disease? Eur. J. Gastroenterol. Hepatol. 11 (Suppl. 1) (1999) S3–S9.
[8] P.J. Kahrilas, R.R. Gupta, The effect of cigarette smoking on salivation and esophageal acid clearance, J. Lab. Clin. Med. 114 (4) (1989) 431–438.
[9] D.A. Katzka, J.E. Pandolfino, P.J. Kahrilas, Phenotypes of gastroesophageal reflux disease: where Rome, Lyon, and Montreal meet, Clin. Gastroenterol. Hepatol. Off. Clin. Pract. J. Am. Gastroenterol. Assoc. 18 (4) (2020) 767–776, https://doi.org/10.1016/j.cgh.2019.07.015.
[10] K.D. Bardhan, V. Strugala, P.W. Dettmar, Reflux revisited: advancing the role of pepsin, Int. J. Otolaryngol. 2012 (2012), 646901, https://doi.org/10.1155/2012/646901.
[11] M.F. Vaezi, S. Singh, J.E. Richter, Role of acid and duodenogastric reflux in esophageal mucosal injury: a review of animal and human studies, Gastroenterology 108 (6) (1995) 1897–1907, https://doi.org/10.1016/0016-5085(95)90156-6.
[12] Q. Zhang, A.T. Agoston, T.H. Pham, et al., Acidic bile salts induce epithelial to mesenchymal transition via VEGF signaling in non-neoplastic Barrett's cells, Gastroenterology 156 (1) (2019) 130–144.e10, https://doi.org/10.1053/j.gastro.2018.09.046.
[13] R.F. Souza, X. Huo, V. Mittal, et al., Gastroesophageal reflux might cause esophagitis through a cytokine-mediated mechanism rather than caustic acid injury, Gastroenterology 137 (5) (2009) 1776–1784, https://doi.org/10.1053/j.gastro.2009.07.055.
[14] D. Castell, J. Murray, R. Tutuian, R. Orlando, R. Arnold, Review article: the pathophysiology of gastro-oesophageal reflux disease-oesophageal manifestations, Aliment. Pharmacol. Ther. 20 (Suppl. 9) (2004) 14–25.
[15] K.B. Dunbar, A.T. Agoston, R.D. Odze, et al., Association of acute gastroesophageal reflux disease with esophageal histologic changes, JAMA 315 (19) (2016) 2104–2112, https://doi.org/10.1001/jama.2016.5657.
[16] A. Muramatsu, T. Azuma, T. Okuda, et al., Association between interleukin-1β-511C/T polymorphism and reflux esophagitis in Japan, J. Gastroenterol. 40 (9) (2005) 873–877, https://doi.org/10.1007/s00535-005-1672-2.
[17] H.H. Cheng, C.S. Chang, H.J. Wang, W.C. Wang, Interleukin-1β and -10 polymorphisms influence erosive reflux esophagitis and gastritis in Taiwanese patients, J. Gastroenterol. Hepatol. 25 (8) (2010) 1443–1451, https://doi.org/10.1111/j.1440-1746.2010.06310.x.
[18] M. Wienbeck, J. Babnert, Epidemiology of reflux disease and reflux esophagitis, Scand. J. Gastroenterol. 24 (Suppl. 156) (1989) 7–13, https://doi.org/10.3109/00365528909091032.
[19] S. Parasa, P. Sharma, Complications of gastro-oesophageal reflux disease, Best Pract. Res. Clin. Gastroenterol. 27 (3) (2013) 433–442, https://doi.org/10.1016/j.bpg.2013.07.002.
[20] J. Labenz, M. Nocon, T. Lind, et al., Prospective follow-up data from the ProGERD study suggest that GERD is not a categorial disease, Off. J. Am. Coll. Gastroenterol. ACG 101 (11) (2006) 2457–2462.
[21] A. Ustaoglu, A. Nguyen, S. Spechler, D. Sifrim, R. Souza, P. Woodland, Mucosal pathogenesis in gastro-esophageal reflux disease, Neurogastroenterol. Motil. Off.

J. Eur. Gastrointest. Motil. Soc. 32 (12) (2020), e14022, https://doi.org/10.1111/nmo.14022.
[22] S.M. D'Souza, K. Houston, L. Keenan, B.S. Yoo, P.J. Parekh, D.A. Johnson, Role of microbial dysbiosis in the pathogenesis of esophageal mucosal disease: a paradigm shift from acid to bacteria? World J. Gastroenterol. 27 (18) (2021) 2054–2072, https://doi.org/10.3748/wjg.v27.i18.2054.
[23] S.M. D'Souza, L.B. Cundra, B.S. Yoo, P.J. Parekh, D.A. Johnson, Microbiome and gastroesophageal disease: pathogenesis and implications for therapy, Ann. Clin. Gastroenterol. Hepatol. 4 (1) (2020) 020–033, https://doi.org/10.29328/journal.acgh.1001018.
[24] Z. Pei, E.J. Bini, L. Yang, M. Zhou, F. Francois, M.J. Blaser, Bacterial biota in the human distal esophagus, Proc. Natl. Acad. Sci. U. S. A. 101 (12) (2004) 4250–4255, https://doi.org/10.1073/pnas.0306398101.
[25] S.A. Fillon, J.K. Harris, B.D. Wagner, et al., Novel device to sample the esophageal microbiome—the esophageal string test, PLoS One 7 (9) (2012), e42938, https://doi.org/10.1371/journal.pone.0042938.
[26] L. Yang, X. Lu, C.W. Nossa, F. Francois, R.M. Peek, Z. Pei, Inflammation and intestinal metaplasia of the distal esophagus are associated with alterations in the microbiome, Gastroenterology 137 (2) (2009) 588–597, https://doi.org/10.1053/j.gastro.2009.04.046.
[27] N. Liu, T. Ando, K. Ishiguro, et al., Characterization of bacterial biota in the distal esophagus of Japanese patients with reflux esophagitis and Barrett's esophagus, BMC Infect. Dis. 13 (2013) 130, https://doi.org/10.1186/1471-2334-13-130.
[28] C.H. Park, S.K. Lee, Exploring esophageal microbiomes in esophageal diseases: a systematic review, J. Neurogastroenterol. Motil. 26 (2) (2020) 171–179, https://doi.org/10.5056/jnm19240.
[29] T.A. Ajayi, S. Cantrell, A. Spann, K.S. Garman, Barrett's esophagus and esophageal cancer: links to microbes and the microbiome, PLoS Pathog. 14 (12) (2018), e1007384, https://doi.org/10.1371/journal.ppat.1007384.
[30] L. Dong, J. Yin, J. Zhao, et al., Microbial similarity and preference for specific sites in healthy oral cavity and esophagus, Front. Microbiol. 9 (2018). https://www.frontiersin.org/article/10.3389/fmicb.2018.01603. (Accessed 17 April 2022).
[31] J.P. Audia, C.C. Webb, J.W. Foster, Breaking through the acid barrier: an orchestrated response to proton stress by enteric bacteria, Int. J. Med. Microbiol. 291 (2) (2001) 97–106, https://doi.org/10.1078/1438-4221-00106.
[32] P. Lund, A. Tramonti, D. De Biase, Coping with low pH: molecular strategies in neutralophilic bacteria, FEMS Microbiol. Rev. 38 (6) (2014) 1091–1125, https://doi.org/10.1111/1574-6976.12076.
[33] M. Begley, C.G.M. Gahan, C. Hill, The interaction between bacteria and bile, FEMS Microbiol. Rev. 29 (4) (2005) 625–651, https://doi.org/10.1016/j.femsre.2004.09.003.
[34] A.D. Brown, H.P. Turner, Membrane stability and salt tolerance in gram-negative bacteria, Nature 199 (4890) (1963) 301–302, https://doi.org/10.1038/199301a0.
[35] H.S. Joo, C.I. Fu, M. Otto, Bacterial strategies of resistance to antimicrobial peptides, Philos. Trans. R. Soc. Lond. Ser. B Biol. Sci. 371 (1695) (2016) 20150292, https://doi.org/10.1098/rstb.2015.0292.
[36] Z. Breijyeh, B. Jubeh, R. Karaman, Resistance of gram-negative bacteria to current antibacterial agents and approaches to resolve it, Molecules 25 (6) (2020) E1340, https://doi.org/10.3390/molecules25061340.
[37] Y. Li, D.A. Powell, S.A. Shaffer, et al., LPS remodeling is an evolved survival strategy for bacteria, Proc. Natl. Acad. Sci. U. S. A. 109 (22) (2012) 8716–8721, https://doi.org/10.1073/pnas.1202908109.

[38] G. Tetz, V. Tetz, Introducing the sporobiota and sporobiome, Gut Pathog. 9 (2017) 38, https://doi.org/10.1186/s13099-017-0187-8.

[39] W.L. Nicholson, N. Munakata, G. Horneck, H.J. Melosh, P. Setlow, Resistance of *Bacillus* endospores to extreme terrestrial and extraterrestrial environments, Microbiol. Mol. Biol. Rev. 64 (3) (2000) 548–572.

[40] I. Okereke, C. Hamilton, G. Reep, et al., Microflora composition in the gastrointestinal tract in patients with Barrett's esophagus, J. Thorac. Dis. 11 (Suppl. 12) (2019) S1581–S1587, https://doi.org/10.21037/jtd.2019.06.15.

[41] J.W. Costerton, Z. Lewandowski, D.E. Caldwell, D.R. Korber, H.M. Lappin-Scott, Microbial biofilms, Annu. Rev. Microbiol. 49 (1) (1995) 711–745, https://doi.org/10.1146/annurev.mi.49.100195.003431.

[42] D.G. Davies, A.M. Chakrabarty, G.G. Geesey, Exopolysaccharide production in biofilms: substratum activation of alginate gene expression by *Pseudomonas aeruginosa*, Appl. Environ. Microbiol. 59 (4) (1993) 1181–1186.

[43] J.W. Costerton, P.S. Stewart, E.P. Greenberg, Bacterial biofilms: a common cause of persistent infections, Science 284 (5418) (1999) 1318–1322, https://doi.org/10.1126/SCIENCE.284.5418.1318.

[44] C. Beloin, A. Roux, J.M. Ghigo, *Escherichia coli* biofilms, Curr. Top. Microbiol. Immunol. 322 (2008) 249–289.

[45] C.M. Dejea, P. Fathi, J.M. Craig, et al., Patients with familial adenomatous polyposis harbor colonic biofilms containing tumorigenic bacteria, Science 359 (6375) (2018) 592–597, https://doi.org/10.1126/science.aah3648.

[46] S. Tomkovich, C.M. Dejea, K. Winglee, et al., Human colon mucosal biofilms from healthy or colon cancer hosts are carcinogenic, J. Clin. Invest. 129 (4) (2019) 1699–1712, https://doi.org/10.1172/JCI124196.

[47] C.M. Dejea, E.C. Wick, E.M. Hechenbleikner, et al., Microbiota organization is a distinct feature of proximal colorectal cancers, Proc. Natl. Acad. Sci. 111 (51) (2014) 18321–18326, https://doi.org/10.1073/pnas.1406199111.

[48] K.L. Blackett, S.S. Siddhi, S. Cleary, et al., Oesophageal bacterial biofilm changes in gastro-oesophageal reflux disease, Barrett's and oesophageal carcinoma: association or causality? Aliment. Pharmacol. Ther. 37 (11) (2013) 1084–1092, https://doi.org/10.1111/apt.12317.

[49] S.H. Rhee, C. Pothoulakis, E.A. Mayer, Principles and clinical implications of the brain–gut–enteric microbiota axis, Nat. Rev. Gastroenterol. Hepatol. 6 (5) (2009), https://doi.org/10.1038/nrgastro.2009.35.

[50] E.A. Mayer, K. Tillisch, A. Gupta, Gut/brain axis and the microbiota, J. Clin. Invest. 125 (3) (2015) 926–938, https://doi.org/10.1172/JCI76304.

[51] M. Camilleri, K. Lasch, W. Zhou, Irritable bowel syndrome: methods, mechanisms, and pathophysiology. The confluence of increased permeability, inflammation, and pain in irritable bowel syndrome, Am. J. Physiol. Gastrointest. Liver Physiol. 303 (7) (2012) G775–G785, https://doi.org/10.1152/ajpgi.00155.2012.

[52] J. Wang, W.D. Chen, Y.D. Wang, The relationship between gut microbiota and inflammatory diseases: the role of macrophages, Front. Microbiol. 11 (2020). https://www.frontiersin.org/article/10.3389/fmicb.2020.01065. (Accessed 24 April 2022).

[53] R. Dheer, R. Santaolalla, J.M. Davies, et al., Intestinal epithelial toll-like receptor 4 signaling affects epithelial function and colonic microbiota and promotes a risk for transmissible colitis, Infect. Immun. 84 (3) (2016) 798–810, https://doi.org/10.1128/IAI.01374-15.

[54] F.A. Amaral, D. Sachs, V.V. Costa, et al., Commensal microbiota is fundamental for the development of inflammatory pain, Proc. Natl. Acad. Sci. U. S. A. 105 (6) (2008) 2193–2197, https://doi.org/10.1073/pnas.0711891105.

[55] T. Patcharatrakul, S. Linlawan, S. Plaidum, S. Gonlachanvit, The effect of rice vs. wheat ingestion on postprandial gastroesophageal reflux (GER) symptoms in patients with overlapping GERD-irritable bowel syndrome (IBS), Foods 11 (1) (2021) 26, https://doi.org/10.3390/foods11010026.
[56] P.B. Mortensen, M.R. Clausen, Short-chain fatty acids in the human colon: relation to gastrointestinal health and disease, Scand. J. Gastroenterol. Suppl. 216 (1996) 132–148, https://doi.org/10.3109/00365529609094568.
[57] R.E. Kuhre, C.F. Deacon, J.J. Holst, N. Petersen, What is an L-cell and how do we study the secretory mechanisms of the L-cell? Front. Endocrinol. 12 (2021), 694284, https://doi.org/10.3389/fendo.2021.694284.
[58] T. Piche, S.B. Des Varannes, S. Sacher-Huvelin, J.J. Holst, J.C. Cuber, J.P. Galmiche, Colonic fermentation influences lower esophageal sphincter function in gastroesophageal reflux disease, Gastroenterology 124 (4) (2003) 894–902, https://doi.org/10.1053/gast.2003.50159.
[59] A. Ropert, C. Cherbut, C. Roze, et al., Colonic fermentation and proximal gastric tone in humans, Gastroenterology 111 (2) (1996) 289–296, https://doi.org/10.1053/gast.1996.v111.pm8690193.
[60] M. El-Salhy, J.G. Hatlebakk, T. Hausken, Possible role of peptide YY (PYY) in the pathophysiology of irritable bowel syndrome (IBS), Neuropeptides 79 (2020), 101973, https://doi.org/10.1016/j.npep.2019.101973.
[61] C. Cherbut, A.C. Aubé, H.M. Blottière, J.P. Galmiche, Effects of short-chain fatty acids on gastrointestinal motility, Scand. J. Gastroenterol. Suppl. 222 (1997) 58–61, https://doi.org/10.1080/00365521.1997.11720720.
[62] T. Piche, Modulation by colonic fermentation of LES function in humans, Am. J. Physiol. Gastrointest. Liver Physiol. 278 (2000) G578–G584.
[63] M.R. Zavala-Solares, G. Fonseca-Camarillo, M. Valdovinos, et al., Gene expression profiling of inflammatory cytokines in esophageal biopsies of different phenotypes of gastroesophageal reflux disease: a cross-sectional study, BMC Gastroenterol. 21 (1) (2021) 201, https://doi.org/10.1186/s12876-021-01707-7.
[64] K. Mönkemüller, T. Wex, D. Kuester, et al., Interleukin-1beta and interleukin-8 expression correlate with the histomorphological changes in esophageal mucosa of patients with erosive and non-erosive reflux disease, Digestion 79 (3) (2009) 186–195, https://doi.org/10.1159/000211714.
[65] A. Cauwels, J. Bultinck, R. De Zwaef, B. Vandendriessche, S. Magez, P. Brouckaert, Nitric oxide production by endotoxin preparations in TLR4-deficient mice, Nitric Oxide 36 (2014) 36–43, https://doi.org/10.1016/j.niox.2013.11.001.
[66] S. Yamato, J.K. Saha, R.K. Goyal, Role of nitric oxide in lower esophageal sphincter relaxation to swallowing, Life Sci. 50 (17) (1992) 1263–1272, https://doi.org/10.1016/0024-3205(92)90326-k.
[67] J.W. Konturek, P. Thor, W. Domschke, Effects of nitric oxide on antral motility and gastric emptying in humans, Eur. J. Gastroenterol. Hepatol. 7 (2) (1995) 97–102.
[68] K. Iijima, T. Shimosegawa, Involvement of luminal nitric oxide in the pathogenesis of the gastroesophageal reflux disease spectrum, J. Gastroenterol. Hepatol. 29 (5) (2014) 898–905, https://doi.org/10.1111/jgh.12548.
[69] N.M. Vaninetti, L. Geldenhuys, G.A. Porter, et al., Inducible nitric oxide synthase, nitrotyrosine and p53 mutations in the molecular pathogenesis of Barrett's esophagus and esophageal adenocarcinoma, Mol. Carcinog. 47 (4) (2008) 275–285, https://doi.org/10.1002/mc.20382.
[70] R. Tomita, K. Tanjoh, S. Fujisaki, M. Fukuzawa, Physiological studies on nitric oxide in the lower esophageal sphincter of patients with reflux esophagitis, Hepato-Gastroenterology 50 (49) (2003) 110–114.

[71] C. Xu, X. Niu, Progress on the mechanism of visceral hypersensitivity in nonerosive reflux disease, Gastroenterol. Res. Pract. 2022 (2022) 4785077, https://doi.org/10.1155/2022/4785077.
[72] E. Deval, X. Gasull, J. Noël, et al., Acid-sensing ion channels (ASICs): pharmacology and implication in pain, Pharmacol. Ther. 128 (3) (2010) 549–558, https://doi.org/10.1016/j.pharmthera.2010.08.006.
[73] R. Cui, L. Zhou, S. Lin, et al., The feasibility of light microscopic measurements of intercellular spaces in squamous epithelium in the lower-esophagus of GERD patients, Dis. Esophagus Off. J. Int. Soc. Dis. Esophagus 24 (1) (2011) 1–5, https://doi.org/10.1111/j.1442-2050.2010.01083.x.
[74] X. Yu, Y. Hu, S. Yu, Effects of acid on vagal nociceptive afferent subtypes in guinea pig esophagus, Am. J. Physiol. Gastrointest. Liver Physiol. 307 (4) (2014) G471–G478,-https://doi.org/10.1152/ajpgi.00156.2014.
[75] R.O. Silva, R.D. Bingana, T.M.A.L. Sales, et al., Role of TRPV1 receptor in inflammation and impairment of esophageal mucosal integrity in a murine model of nonerosive reflux disease, Neurogastroenterol. Motil. Off. J. Eur. Gastrointest. Motil. Soc. (2018) e13340. Published online March 23, 2018 https://doi.org/10.1111/nmo.13340.
[76] N. Yoshida, M. Kuroda, T. Suzuki, et al., Role of nociceptors/neuropeptides in the pathogenesis of visceral hypersensitivity of nonerosive reflux disease, Dig. Dis. Sci. 58 (8) (2013) 2237–2243, https://doi.org/10.1007/s10620-012-2337-7.

CHAPTER 5

Gastroesophageal reflux disease and the role of *Helicobacter pylori*

Michelle Eugene Lee and Philip O. Katz
Department of Medicine, Division of Gastroenterology and Hepatology, New York Presbyterian—Weill Cornell, New York, NY, United States

Introduction

Gastroesophageal reflux disease (GERD) continues to be among the most common conditions viewed by gastroenterologists, surgeons, and primary care practitioners [1]. Although GERD is associated with or affected by many other diseases and clinical conditions, "proving" these associations is often difficult, both because of the widespread prevalence of GERD as well as the lack of direct evidence to confirm cause and effect. *Helicobacter pylori* infection is one of those clinical conditions. The organism is a gram-negative bacterium that when it infects humans, resides exclusively in the gastric mucosa, and is an accepted etiology of peptic ulcer disease and a major risk factor for gastric cancer [2]. As efforts to eradicate the organism from anyone infected progressed, clinical and epidemiologic observations have suggested that eradication of *H. pylori* might influence exacerbation or development of GERD [3,4]. These reports raise controversy as to how the infections might be managed in the setting of GERD. In this chapter, we address the proposed and postulated relationships between GERD and *H. pylori* to put them under clinical perspective.

Pathogenesis of GERD

There are multiple pathophysiologic features of GERD. The primary one relates to abnormal function of the esophagogastric junction—specifically, the antireflux barrier which is composed of the lower esophageal sphincter (LES) and crural diaphragm [5]. Abnormal function of the LES characterized by either transient inappropriate relaxations or a constantly reduced pressure, combined with an anatomic change in the crura, results in weakening and subsequent development of a hiatal hernia. These changes create a milieu for

Esophageal Disease and the Role of the Microbiome
https://doi.org/10.1016/B978-0-323-95070-1.00007-2

the retrograde reflux of gastric contents into the esophagus. Impaired esophageal clearance as well as prolonged contact with gastric refluxate and consequent reductions of esophageal mucosal integrity led to symptomatic endoscopic or histologic evidence of injury. Other contributors to GERD are decreased salivary production, delayed gastric emptying, and abnormal esophageal sensitivity. Symptoms and esophageal mucosal injury are ultimately caused by gastric refluxate, which is principally acid. Any alteration in gastric acid secretion and subsequently the gastric refluxate, in the setting of a defective esophagogastric junction, has the potential to alter symptoms and/or incur esophageal injury [6].

H. pylori is a spiral-shaped gram-negative microaerophilic bacterium which, when evident, typically colonizes the stomach of humans [7,8]. Infection is present with variable frequency across geographic locations [9]. Although there is a recognizable decrease in prevalence in some parts of the world, it appears that close to half of the world's population is currently infected [10]. In 1983, Warren and Marshall demonstrated a direct connection between *H. pylori* and antral gastritis, as well as gastric and duodenal ulcers [11]. Marshall ingested the bacteria and developed acute gastritis, fulfilling Koch's postulate and cementing a cause-and-effect relationship between *H. pylori* and gastritis [7,12]. Since that experiment, *H. pylori* colonization of the stomach has been widely accepted as the main cause of antral gastritis and peptic ulcer disease [2]. This is exclusive of nonsteroidal antiinflammatory drugs which are more likely to cause injury in the setting of *H. pylori* infection [13]. Classified by the World Health Organization as a class 1 (gastric) carcinogen, the organism is a major risk factor for gastric adenocarcinoma and MALT lymphoma [14,15]. Extensive literature offers a variable but accepted relationship between *H. pylori* and the symptom of dyspepsia [16]. The organism resides exclusively in gastric tissue, initially in the antrum where there are no acid-producing parietal cells [17]. Through downstream effects from here, acid production may be increased. If not detected or eradicated, *H. pylori* may proliferate, eventually spreading to the corpus and resulting in the development of gastric atrophy, creating the milieu which likely increases the risk for the development of noncardia gastric adenocarcinoma [17]. As *H. pylori* has no proven direct effect on LES function, hiatal hernia development, or other proven GERD-related physiologic abnormality, the organism's possible alteration of acid production is believed to potentially effect GERD at this time. However, it is possible that *H. pylori* infection may affect other nongastric/gut microbiome that could also influence these pathogenic factors [18–20].

Based on the variability of gastric infection and the effects of eradication of *H. pylori*, gastric acidity may either be increased or decreased. Eradication may result in potentially opposite effects on GERD—causing either an increase in acidity to make GERD symptoms or injury more likely, or paradoxically decreasing acidity which may protect against GERD, Barrett's esophagus, or esophageal adenocarcinoma.

Biologic links/plausibility

H. pylori causes different patterns of gastritis, which have potentially opposite effects on acid secretion and thereby related GERD. The most common, antral-predominant gastritis, interferes with the feedback loop of parietal cell secretion, which is associated with increased acid secretion and duodenal ulcer disease. In this situation, GERD might be worsened [21]. Corpus-predominant or pangastritis causes local inflammation, resulting in reduced or no acid secretion if atrophy occurs [22]. This atrophic gastritis is associated with an increased risk of gastric cancer [17]. In this setting, protection against erosive esophagitis and Barrett's esophagus may be possible. A summary of these findings is shown in Fig. 1.

Antral-predominant gastritis alters the hormonal effects of gastrin and somatostatin, thereby increasing acid secretion and worsening GERD [22]. Gastrin is secreted from G cells in the antrum, which stimulates parietal cells to release gastric acid both directly on cholecystokinin 2 (CCK2) and indirectly via histamine on CCK2 [21]. The parietal cells exist in the corpus of the stomach. Somatostatin inhibits gastric acid secretion, most likely through inhibition of histamine secretion, and is released from the D cell present throughout the gastrointestinal tract [22]. The regulation of gastric acid depends on the interplay of gastrin and somatostatin, which will ultimately have the potential to affect GERD [23]. Notably, the antrum is predominantly composed of gastrin cells, and few to none exist in the corpus [23].

H. pylori has been found to disrupt these hormonal controls of gastric acid. Liu et al. evaluated patients with dyspepsia but not reflux esophagitis, peptic ulcers, or malignancy, via endoscopic biopsies of the corpus and antrum which were histologically evaluated for G and D cells [23]. In this cohort, *H. pylori* positive patients displayed antral-predominant inflammation and atrophy. *H. pylori* infected patients had significantly more immunoreactive G cells and fewer immunoreactive D cells than both those with noninfected gastritis and those with histological normal gastric mucosa

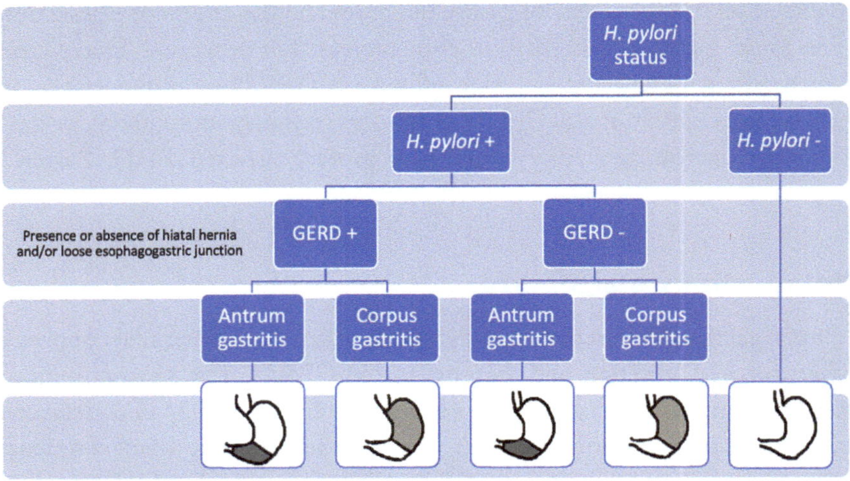

Acid secretion	↑ to normal	↓ to normal	↑ to normal	↓ to normal	normal
Neoplastic risk		2-3x increased risk gastric cancer		2-3x increased risk gastric cancer	8-20x increased risk esophageal cancer
Ulcer risk	10% lifetime risk duodenal ulcer		10% lifetime risk duodenal ulcer		
H pylori eradication	GERD may improve Decreased duodenal ulcer	GERD may worsen Decreased PPI efficacy	GERD should not develop Decreased duodenal ulcer	Unmask subclinical GERD PPI decreased efficacy	

Fig. 1 The distribution of gastritis caused by *H. pylori* in the stomach influences its effect on GERD. Antral gastritis induces more acid secretion and is a risk factor for duodenal ulcers, while corpus gastritis decreases acid in the stomach and carries a greater risk of gastric cancer. *GERD*, gastroesophageal reflux disease.

(70% G:5% D in *H. pylori*+, 47% G:15% D in *H. pylori*−, $P<.01$). Other studies have also shown that antral gastritis from *H. pylori* is associated with lower somatostatin and higher gastrin levels [24–27]. Cytokines released in gastritis, such as tumor necrosis factor alpha, as well as byproducts of *H. pylori*, such as ammonia, are thought to stimulate gastrin release from G cells [22]. Ultimately antral gastritis is likely to result in an increase in acid production which may explain the finding that the Western hemisphere, with predominantly antral *H. pylori* gastritis, currently has higher rates of GERD [9].

In contrast, multifocal *H. pylori* infection extending to the corpus and/or fundus of the stomach causes a decrease in acid secretion and may protect against GERD. When *H. pylori* acts on the parietal cells, interleukin 1 beta from *H. pylori* inhibits the parietal cells and histamine release from enterochromaffin-like cells and promotes atrophy of the area, killing parietal cells [22]. The resulting clinical finding is atrophic gastritis, which is most common in China and Japan but not uncommon in the West [28].

Several Chinese and Japanese studies have demonstrated a prevalence of atrophic gastritis greater than 50%, while two studies from Linqu County, China had almost universal chronic atrophic gastritis [29–31]. A well-done case-control study from the Houston Veterans Affairs hospital found that corpus gastritis and atrophy were inversely associated with Barrett's esophagus [32].

In addition, it has been shown that atrophic gastritis is associated with a higher prevalence of cytotoxin-associated gene (*cagA*) strains of *H. pylori*. A toxin produced by the *cagA* gene changes the gastric lining cell structure and facilitates *H. pylori* attachment [33]. The *cagA* gene has been identified in about 50%–60% of *H. pylori* infections from Western countries and >90% of isolates from Asian countries [34]. Studies in North America, Europe, Asia, and Australia demonstrate that infection with *cagA*-positive strains carries a decreased risk of GERD complications including erosive esophagitis and Barrett's esophagus [32,35–38]. A meta-analysis in 2018 examined 72 studies worldwide and found that the risk reduction of Barrett's esophagus with *H. pylori* infection was persistent and independent of geographic location [39]. The Barrett's and Esophageal Adenocarcinoma Consortium (BEACON) case-control study concluded that this effect is probably mediated by a decrease in GERD in *H. pylori* infected patients, since the protective effect against Barrett's disappears in patients with GERD symptoms [40].

A variety of other physiologic changes may also play a role in decreasing acid secretion and GERD in *H. pylori* infected patients. *H. pylori* infection increases gastrin levels. Gastrin has been shown to stimulate lower esophageal sphincter contraction and thereby enhance the gastroesophageal reflux barrier. While not directly studied, this effect may reduce the frequency of transient LES relaxations [41]. In addition, ghrelin, an orexigenic hormone that stimulates gastric acid secretion, may be lower in patients with *H. pylori* than those without, thereby inducing a decrease in esophageal symptoms or injury [42,43]. Furthermore, alkaline ammonia produced by *H. pylori* in the cardia may protect esophageal tissue from postprandial acid [44]. All these potential direct or indirect changes in gastric acidity induced by *H. pylori* infection, and the pattern of gastritis produced by the infection, give credence to the methods in which *H. pylori* can impact GERD.

Epidemiology/outcomes

A substantial portion of the world's population has been infected with *H. pylori*, ranging widely between developed and underdeveloped nations.

This is illustrated by a meta-analysis that estimated that *H. pylori* prevalence ranged from a high of 88% in Nigeria to a low of 19% prevalence in Switzerland [9]. *H. pylori* infection is commonly acquired in childhood, usually, before the age of 12 [45]. A major predictor of the prevalence of *H. pylori* infection in a region is socioeconomic status during childhood. The infection is more prevalent in areas of lower socioeconomic status which historically reflects the density of living, educational level, hygiene, and sanitation [46]. Though overall there appears to be a parallel in the relationship between the decrease of *H. pylori* infection and the increase in GERD prevalence, the prevalence of GERD cannot always be predicted by the prevalence of *H. pylori* infection as one study from Nigeria estimated a GERD symptom prevalence as high as 32% [47,48].

Studies performed using the United States Veterans Administration database demonstrated a relationship between the decline in two *H. pylori* related diseases, peptic ulcer disease and noncardia gastric adenocarcinoma, and a precipitous increase in the prevalence of GERD and GERD complications [49]. Although the prevalence of *H. pylori* was no difference between patients with esophagitis and controls, the pattern of gastritis appeared to be important as corpus gastritis was associated with a 54% reduced risk of esophagitis (95% CI, 21%–73%) [50].

A Chilean study demonstrated a significant increase in GERD among patients with *H. pylori*, while North American and Asian countries had significantly less GERD in patients with *H. pylori* [51]. Subsequent large studies from Asia also found that the presence of *H. pylori* infection was associated with a decreased risk of erosive esophagitis [52,53]. A case-control study from Korea of over 5600 patients undergoing routine endoscopy for gastric cancer screening demonstrated a reduced prevalence of *H. pylori* infection in patients with reflux esophagitis than in those without [3]. A Chinese study of 375 patients showed that the prevalence of *H. pylori* infection decreased stepwise from reflux esophagitis grade I (51.72%), grade II (28.57%), and then to grade III (20.68%) ($P < .05$) [54]. These studies do not specifically address the frequency of corpus or atrophic gastritis, but do come from areas in which one would suspect a high prevalence of this pattern of infection. In contrast, a population-based study of 65,000 people in Norway and an American study of 533 men assessed GERD symptoms and found no relationship with *H. pylori* infection, regardless of *cag-A* status [37,55]. Neither endoscopic esophagitis nor 24 hour pH monitoring was significantly associated with *H. pylori* infection in a 100-patient study from Spain. While no causal link between *H. pylori* and GERD can be proven, the overall the balance of studies from the United States and large-population studies in Asia seem to favor *H. pylori* infection as protective against severe endoscopic reflux esophagitis [4,35,56–61].

Given the potential effects of *H. pylori* infection and GERD, including complications and sequelae, questions have been raised regarding the effect of *H. pylori* eradication on exacerbation or precipitation of new onset GERD. In addition, questions have been raised asking if or how *H. pylori* infection might affect treatment decisions in patients with GERD. One might surmise that if antral gastritis is the predominant pattern, eradication should result in an overall reduction in both duodenal ulcer disease and gastric cancer, and overall have minimal to no major effects on GERD. In areas where corpus gastritis and atrophy predominate, we might "trade" a reduction in ulcers and gastric cancer with an increase in gastric acid secretion, more GERD, and ultimately GERD complications. Teleologically, one would hypothesize that any increase or rebound in acid secretion should not be permanent thus any increase in GERD symptoms would be transient. There are several proposed mechanisms by which *H. pylori* eradication and or treatment could influence GERD symptoms. The ammonia produced by *H. pylori* has been hypothesized to augment the acid-suppressive effects of PPI treatment [62]. Additionally, PPIs may change *H. pylori* gastritis into a corpus-predominant gastritis, inhibiting parietal cells and reducing acid secretion [63]. It is well known that the increase in worldwide obesity, with increased BMI and predominantly waist circumference, has impacted GERD. The hormone ghrelin appears to be important in this relationship. *H. pylori* eradication has been shown to result in increased levels of ghrelin production which increases gastric acid secretion and stimulates appetite [64–66]. However, overall studies on the relationship between *H. pylori* elimination and increased body mass index (BMI) and de-novo GERD have demonstrated mixed results. A randomized control trial of 1558 *H. pylori*-positive patients in Europe showed that BMI increased, and dyspepsia decreased in patients who received eradication treatment instead of placebo [67]. This study used a histamine H2-receptor antagonist rather than PPI for *H. pylori* treatment. Conversely, a randomized controlled trial followed 404 *H. pylori*-positive patients for 12 months and found that *H. pylori* treatment did not predict new onset GERD, although increased BMI was associated with new esophagitis on endoscopy [64,65].

In addition to an unclear mechanism of action, some studies on *H. pylori* treatment indicate that there are protective effects of *H. pylori* against GERD while larger meta-analyses and systematic reviews demonstrate no association. Labenz et al. followed 244 German duodenal ulcer patients for 3 years after treatment of *H. pylori* and found a significantly increased frequency of erosive esophagitis on endoscopy in the successfully treated group compared to 214 patients with ongoing infection (25.8%–12.9%, $P < .001$). Patients who developed esophagitis had more severe corpus gastritis before treatment and more frequently gained weight [68]. Another study from the same country followed 846 patients with endoscopic esophagitis treated with a PPI who were reassessed endoscopically at 4 or 8 weeks. In patients with *H. pylori*, the esophagitis healing rates were significantly higher after 4 (86.6%–76.3%; $P = .0005$) and 8 weeks (96.4%–91.8%; $P < .004$). Patients with *H. pylori* also experienced significantly greater GERD symptom relief after 4 weeks than those uninfected [69]. While statistically significant the overall difference does not appear clinically important and long term follow up was not available. Several studies from Asia support that patients may develop GERD after *H. pylori* eradication. In a cohort of 10,102 asymptomatic individuals in Korea, the prevalence of reflux esophagitis was significantly greater in patients with *H. pylori* than without (6.4%–3.3%, $P < .001$). Furthermore, the odds ratio for the presence of esophagitis was 2.34 (95% CI, 1.45–3.76; $P < .001$) in patients with successful *H. pylori* eradication compared to those with persistent infection. The overall prevalence of esophagitis is so low that while supportive of the possibility of new GERD it is of no clinical importance especially as treatment did not influence GERD symptoms [70]. In a Japanese study patients with *H. pylori* eradication and presence of hiatal hernia were more likely to have esophagitis than those with failed treatment, though overall numbers are low [71].

Conflicting evidence comes from meta-analyses and systematic reviews showing that *H. pylori* treatment resulted in no significant change in GERD symptoms or rates of esophagitis. A meta-analysis of 16 cohort studies of 6404 patients and found that *H. pylori* eradication did not affect the development of erosive esophagitis or GERD symptoms (OR 0.87; 95% CI, 0.66–1.14) [72]. The study demonstrated limited heterogeneity with no differences detected even when studies were subgrouped by region, age, and baseline presence of GERD, peptic ulcer, or dyspepsia. A systematic review of 11 randomized controlled trials of patients with peptic ulcer disease or dyspepsia, and found no significant difference in the development of GERD symptoms or esophagitis based on *H. pylori* treatment status, including when analyzed by country of origin [73].

While epidemiologic findings suggest that *H. pylori* in certain geographic regions protects against GERD, clinical studies have not consistently shown that *H. pylori* has anything more than a short-term effect on the development or improvement of GERD symptoms or esophagitis. In Delaney and McColl's review of the topic, they propose possible reasons for this discrepancy [74]. Because the protective nature of *H. pylori* against GERD depends on the distribution of gastritis within the stomach, they propose that different results can be a result of the percentages of antral versus corpus predominant gastritis represented in a country or study population. Furthermore, epidemiologic studies are vulnerable to confounding factors. Direct prospective randomized control trials have greater control over confounding factors, but outcomes such as GERD symptom questionnaires are less subjective than findings such as erosive esophagitis on endoscopy. Other variables that can also impact the influence of *H. pylori* infection on GERD include age, socioeconomic status, host genetic factors, and virulence factors such as *cagA* status. As almost all studies are retrospective in nature, methodologic issues will always confound the results.

The most recent international recommendations were published in 2021 by the World Gastroenterology Organization on *H. pylori* diagnosis and treatment, and reflect the variable epidemiologic findings globally. These guidelines suggested that patients with *H. pylori* and GERD on long-term PPI be treated for *H. pylori*, with the caveat that the strength of these recommendations may vary regionally [46]. Additionally, the Maastricht V/Florence Consensus Report in 2017 gathered 43 experts from 24 countries to review the existing *H. pylori* literature using the Delphi process [75]. Because *H. pylori* gastritis has variable effects on acid secretion based on the pattern of infection, eradication can partially correct the either higher or lower acid state. They concluded that observed changes in acid production after *H. pylori* treatment have no proven clinical relevance in most populations, and thus the decision to treat *H. pylori* should be made independent of the presence of GERD symptoms. Because *H. pylori* is a major risk factor for gastric cancer, gastric MALT lymphoma, and peptic ulcer disease, the 2015 Kyoto Consensus recommended to treat all patients who are found to be *H. pylori* positive [45].

Within the United States, the American College of Gastroenterology recommended in 2017 that patients with symptoms of GERD and no history of peptic ulcer disease should not be routinely tested for *H. pylori*, as no clear benefit of treatment has been shown. However, if testing is performed and it is positive, *H. pylori* treatment should be offered [76]. Similar to the World Gastroenterology Organization, the Houston Consensus Conference in 2018 concluded that patients receiving long-term PPIs (over 1 month) should be tested and treated for *H. pylori*, given the risk of atrophic gastritis leading to metaplasia and/or gastric carcinoma. Additionally, they recommended that mainly GERD patients at high risk for *H. pylori*, such as those of Latino or African American ethnicity, should be routinely tested [77]. Table 1 summarizes these various published clinical recommendations.

Table 1 Guideline-forming meetings in the last several years have provided various clinical recommendations on *H. pylori* testing and treatment in the context of GERD.

	Method and strength of recommendations	Recommendations
Kyoto Global Consensus Report (2015)	*Delphi* Grade of recommendation: strong Evidence level: high Consensus level: 100%	*H. pylori* infected individuals should be offered eradication therapy unless there are competing considerations.
American College of Gastroenterology Guidelines (2017)	*GRADE* Strong recommendation High quality of evidence	Patients with typical symptoms of GERD who do not have a history of peptic ulcer disease need not be tested for *H. pylori* infection. Patients found to be infected should be offered eradication treatment.
Maastricht V/Florence Consensus Report (2017)	*Delphi* Grade of recommendation: high Evidence level: high	*H. pylori* gastritis may increase or decrease acid secretion. Treatment may reverse or partially reverse these effects.
Houston Consensus Conference (2018)	*GRADE* Grade 1C 91% agree/strongly agree	Routine testing for *H. pylori* infection in patients with reflux symptoms only if they are at high risk for *H. pylori*-related disease.
	Grade 2C 82% agree/strongly agree	Patients receiving long-term PPIs (>1 month) should be tested for *H. pylori* infection.
World Gastroenterology Organization (2021)	*Expert consensus* Strength of indications may vary regionally and individually	Recommend treatment of *H. pylori* positive patients with GERD who require long-term PPI

Conclusions

The interplay of *H. pylori* and GERD has captured the interest of gastroenterologists and bacteriologists, with proposals that *H. pylori* may act as a commensal or symbiotic organism instead of a unilateral pathogen. The pattern of gastritis produced by *H. pylori* likely determines much of the relationship of the bacteria in GERD and its complications. Epidemiologic data support that eradication of *H. pylori* from the planet has played perhaps some role in the increase in GERD. However, the widespread increase in obesity, changes in diet, and other sociologic issues are likely more important than *H. pylori* eradication of GERD itself. While the epidemiologic data suggest a possible protective effect of corpus-predominant *H. pylori* gastritis against complications of GERD, in particular Barrett's esophagus, studies on outcomes of eradication on these complications are less consistently conclusive or not available. In concordance with most expert consensuses, we support the treatment of *H. pylori* if present regardless of the presence of GERD or the unlikely possibility of short-term exacerbation of GERD symptoms. No evidence exists that eradication will induce GERD long-term. In addition, patients who are already on long-term PPI should be treated for *H. pylori* if present. The benefit of reducing complications of peptic ulcer disease and gastric cancer is well accepted. Routine testing for *H. pylori* in patients with GERD and avoidance of eradication for concerns regarding GERD cannot be supported.

Future directions

There is no momentum to do anything to change the clinical recommendation to eradicate *H. pylori* whenever an infection is documented. The relationship between *H. pylori* and GERD is complex and controversial. Opportunities to evaluate the long term effects of GERD are few and at the time of this writing would all be retrospective. Patients with esophageal adenocarcinoma might be more carefully studied, making attempts to determine if *H. pylori* was ever present and/or treated. Patients with Barrett's esophagus could be assessed for baseline prevalence and followed for progression. All these studies are likely impractical. Of interest is the development of potassium competitive acid blockers (P-CAB) in both Asia and the United States. In the United States one such medication, Vonoprazan, has been studied and just approved for eradication of *H. pylori*. Perhaps these patients can be assessed for the presence of new onset GERD symptoms or erosive esophagitis if endoscopic reports are available. The background prevalence of Barrett's esophagus in these populations might then be compared to historical controls and compared with PPI outcome data.

Conflict of interest disclosure

Philip O. Katz: Consultant Phathom Pharma; Michelle Eugene Lee: None.

References

[1] S.M. Ornstein, P.J. Nietert, R.G. Jenkins, C.B. Litvin, The prevalence of chronic diseases and multimorbidity in primary care practice: a PPRNet report, J. Am. Board Fam. Med. 26 (5) (2013) 518–524, https://doi.org/10.3122/jabfm.2013.05.130012.

[2] M. Dixon, *Helicobacter pylori*: physiology and genetics, in: Pathology of Gastritis and Peptic Ulceration, ASM Press, 2001 (Chapter 38).

[3] S.J. Chung, S.H. Lim, J. Choi, et al., *Helicobacter pylori* serology inversely correlated with the risk and severity of reflux esophagitis in *Helicobacter pylori* endemic area: a matched case-control study of 5,616 health check-up Koreans, J. Neurogastroenterol. Motil. 17 (3) (2011) 267–273, https://doi.org/10.5056/jnm.2011.17.3.267.

[4] J.C. Wu, J.J. Sung, E.K. Ng, et al., Prevalence and distribution of *Helicobacter pylori* in gastroesophageal reflux disease: a study from the East, Am. J. Gastroenterol. 94 (7) (1999) 1790–1794, https://doi.org/10.1111/j.1572-0241.1999.01207.x.

[5] H. Mousa, M. Hassan, Gastroesophageal reflux disease, Pediatr. Clin. N. Am. 64 (3) (2017) 487–505, https://doi.org/10.1016/j.pcl.2017.01.003.

[6] P.O. Katz, K.B. Dunbar, F.H. Schnoll-Sussman, K.B. Greer, R. Yadlapati, S.J. Spechler, ACG clinical guideline for the diagnosis and management of gastroesophageal reflux disease, Am. J. Gastroenterol. 117 (1) (2022) 27–56, https://doi.org/10.14309/ajg.0000000000001538.

[7] B.J. Egan, C.A. O'Morain, A historical perspective of *Helicobacter* gastroduodenitis and its complications, Best Pract. Res. Clin. Gastroenterol. 21 (2) (2007) 335–346, https://doi.org/10.1016/j.bpg.2006.12.002.

[8] G. Bizzozero, Ueber die schlauchförmigen Drüsen des Magendarmkanals und die Beziehungen ihres Epithels zu dem Oberflächenepithel der Schleimhaut Dritte Mittheilung, Arch. Mikrosk. Anat. Enwicklmech. 42 (1) (1893) 82–152, https://doi.org/10.1007/BF02975307.

[9] J.K.Y. Hooi, W.Y. Lai, W.K. Ng, et al., Global prevalence of *Helicobacter pylori* infection: systematic review and meta-analysis, Gastroenterology 153 (2) (2017) 420–429, https://doi.org/10.1053/j.gastro.2017.04.022.

[10] B.A. Salih, *Helicobacter pylori* infection in developing countries: the burden for how long? Saudi J. Gastroenterol. 15 (3) (2009) 201–207, https://doi.org/10.4103/1319-3767.54743.

[11] B. Marshall, J.R. Warren, Unidentified curved bacilli in the stomach of patients with gastritis and peprtic ulceration, Lancet 323 (8390) (1984) 1311–1315, https://doi.org/10.1016/S0140-6736(84)91816-6.

[12] J.R. Warren, The discovery of *Helicobacter pylori* in Perth, Western Australia, in: Helicobacter Pioneers (Firsthand Account from the Scientists Who Discovered Helicobacters 1892–1982), Blackwell Science, Asia, 2002, pp. 151–164.

[13] D.M. McCarthy, *Helicobacter pylori* and non-steroidal anti-inflammatory drugs: does infection affect the outcome of NSAID therapy? Yale J. Biol. Med. 71 (2) (1998) 101–111.

[14] Schistosomes, liver flukes and *Helicobacter pylori*. IARC Working Group on the evaluation of carcinogenic risks to humans. Lyon, 7–14 June 1994, IARC Monogr. Eval. Carcinog. Risks Hum. 61 (1994) 1–241.
[15] W. Zhang, H. Lu, D.Y. Graham, An update on *Helicobacter pylori* as the cause of gastric cancer, Gastrointest. Tumors 1 (3) (2014) 155–165, https://doi.org/10.1159/000365310.
[16] H. Suzuki, P. Moayyedi, *Helicobacter pylori* infection in functional dyspepsia, Nat. Rev. Gastroenterol. Hepatol. 10 (3) (2013) 168–174, https://doi.org/10.1038/nrgastro.2013.9.
[17] G. Carrasco, A.H. Corvalan, *Helicobacter pylori*-induced chronic gastritis and assessing risks for gastric cancer, Gastroenterol. Res. Pract. 2013 (2013) 393015, https://doi.org/10.1155/2013/393015.
[18] L. Yang, F. Francois, Z. Pei, Molecular pathways: pathogenesis and clinical implications of microbiome alteration in esophagitis and Barrett esophagus, Clin. Cancer Res. 18 (8) (2012) 2138–2144, https://doi.org/10.1158/1078-0432.ccr-11-0934.
[19] K.L. Blackett, S.S. Siddhi, S. Cleary, et al., Oesophageal bacterial biofilm changes in gastro-oesophageal reflux disease, Barrett's and oesophageal carcinoma: association or causality? Aliment. Pharmacol. Ther. 37 (11) (2013) 1084–1092, https://doi.org/10.1111/apt.12317.
[20] S.M. D'Souza, L. Cundra, B.S. Yoo, P.J. Parekh, D.A. Johnson, Microbiome and gastroesophageal disease: pathogenesis and implications for therapy, Ann. Clin. Gastroenterol. Hepatol. (2020) 020–033.
[21] E. Bitziou, B.A. Patel, Simultaneous detection of gastric acid and histamine release to unravel the regulation of acid secretion from the guinea pig stomach, Am. J. Physiol. Gastrointest. Liver Physiol. 303 (3) (2012) G396–G403, https://doi.org/10.1152/ajpgi.00548.2011.
[22] J. Calam, A. Gibbons, Z.V. Healey, P. Bliss, N. Arebi, How does *Helicobacter pylori* cause mucosal damage? Its effect on acid and gastrin physiology, Gastroenterology 113 (6 Suppl) (1997) S43–S49. discussion S50 https://doi.org/10.1016/s0016-5085(97)80010-8.
[23] Y. Liu, G.D.C. Vosmaer, G.N.J. Tytgat, S.D. Xiao, F.J.W. Ten Kate, Gastrin (G) cells and somatostatin (D) cells in patients with dyspeptic symptoms: *Helicobacter pylori* associated and non-associated gastritis, J. Clin. Pathol. 58 (9) (2005) 927–931, https://doi.org/10.1136/jcp.2003.010710.
[24] A.H. Gibbons, S. Legon, M.M. Walker, M. Ghatei, J. Calam, The effect of gastrin-releasing peptide on gastrin and somatostatin messenger RNAs in humans infected with *Helicobacter pylori*, Gastroenterology 112 (6) (1997) 1940–1947, https://doi.org/10.1053/gast.1997.v112.pm9178686.
[25] M. Sumii, K. Sumii, A. Tari, et al., Expression of antral gastrin and somatostatin mRNA in *Helicobacter pylori*-infected subjects, Am. J. Gastroenterol. 89 (9) (1994) 1515–1519.
[26] H. Kaneko, K. Nakada, T. Mitsuma, et al., *Helicobacter pylori* infection induces a decrease in immunoreactive-somatostatin concentrations of human stomach, Dig. Dis. Sci. 37 (3) (1992) 409–416, https://doi.org/10.1007/bf01307736.
[27] L. Odum, H.D. Petersen, I.B. Andersen, B.F. Hansen, J.F. Rehfeld, Gastrin and somatostatin in *Helicobacter pylori* infected antral mucosa, Gut 35 (5) (1994) 615–618, https://doi.org/10.1136/gut.35.5.615.
[28] M.N. Weck, H. Brenner, Prevalence of chronic atrophic gastritis in different parts of the world, Cancer Epidemiol. Biomarkers Prev. 15 (6) (2006) 1083–1094, https://doi.org/10.1158/1055-9965.EPI-05-0931.
[29] W.C. You, W.J. Blot, J.Y. Li, et al., Precancerous gastric lesions in a population at high risk of stomach cancer, Cancer Res. 53 (6) (1993) 1317–1321.
[30] M. Asaka, M. Kato, M. Kudo, et al., Atrophic changes of gastric mucosa are caused by *Helicobacter pylori* infection rather than aging: studies in asymptomatic Japanese adults, Helicobacter 1 (1) (1996) 52–56, https://doi.org/10.1111/j.1523-5378.1996.tb00008.x.

[31] W.C. You, L. Zhang, M.H. Gail, et al., Precancerous lesions in two counties of China with contrasting gastric cancer risk, Int. J. Epidemiol. 27 (1998) 945–948.
[32] L.A. Fischbach, D.Y. Graham, J.R. Kramer, et al., Association between *Helicobacter pylori* and Barrett's esophagus: a case-control study, Am. J. Gastroenterol. 109 (3) (2014) 357–368, https://doi.org/10.1038/ajg.2013.443.
[33] F. Bagnoli, L. Buti, L. Tompkins, A. Covacci, M.R. Amieva, *Helicobacter pylori* CagA induces a transition from polarized to invasive phenotypes in MDCK cells, Proc. Natl. Acad. Sci. U. S. A. 102 (45) (2005) 16339–16344, https://doi.org/10.1073/pnas.0502598102.
[34] P. Correa, M.B. Piazuelo, Natural history of *Helicobacter pylori* infection, Dig. Liver Dis. 40 (7) (2008) 490–496, https://doi.org/10.1016/j.dld.2008.02.035.
[35] M.F. Vaezi, G.W. Falk, R.M. Peek, et al., CagA-positive strains of *Helicobacter pylori* may protect against Barrett's esophagus, Am. J. Gastroenterol. 95 (9) (2000) 2206–2211, https://doi.org/10.1111/j.1572-0241.2000.02305.x.
[36] J.J. Vicari, R.M. Peek, G.W. Falk, et al., The seroprevalence of cagA-positive *Helicobacter pylori* strains in the spectrum of gastroesophageal reflux disease, Gastroenterology 115 (1) (1998) 50–57, https://doi.org/10.1016/s0016-5085(98)70364-6.
[37] J.H. Rubenstein, J.M. Inadomi, J. Scheiman, et al., Association between *Helicobacter pylori* and Barrett's esophagus, erosive esophagitis, and gastroesophageal reflux symptoms, Clin. Gastroenterol. Hepatol. 12 (2) (2014) 239–245, https://doi.org/10.1016/j.cgh.2013.08.029.
[38] D.A. Corley, A. Kubo, T.R. Levin, et al., *Helicobacter pylori* infection and the risk of Barrett's oesophagus: a community-based study, Gut 57 (6) (2008) 727–733, https://doi.org/10.1136/gut.2007.132068.
[39] B. Erőss, N. Farkas, Á. Vincze, et al., *Helicobacter pylori* infection reduces the risk of Barrett's esophagus: a meta-analysis and systematic review, Helicobacter 23 (4) (2018), e12504, https://doi.org/10.1111/hel.12504.
[40] Z. Wang, N.J. Shaheen, D.C. Whiteman, et al., *Helicobacter pylori* infection is associated with reduced risk of Barrett's esophagus: an analysis of the Barrett's and Esophageal Adenocarcinoma Consortium, Am. J. Gastroenterol. 113 (8) (2018) 1148–1155, https://doi.org/10.1038/s41395-018-0070-3.
[41] D.O. Castell, L.D. Harris, Hormonal control of gastroesophageal-sphincter strength, N. Engl. J. Med. 282 (16) (1970) 886–889, https://doi.org/10.1056/NEJM197004162821602.
[42] P. Mantero, G.S. Matus, R.E. Corti, et al., *Helicobacter pylori* and corpus gastric pathology are associated with lower serum ghrelin, World J. Gastroenterol. 24 (3) (2018) 397–407, https://doi.org/10.3748/wjg.v24.i3.397.
[43] C.V. Nweneka, A.M. Prentice, *Helicobacter pylori* infection and circulating ghrelin levels—a systematic review, BMC Gastroenterol. 11 (2011) 7, https://doi.org/10.1186/1471-230x-11-7.
[44] K.E.L. McColl, H. Watabe, M.H. Derakhshan, Role of gastric atrophy in mediating negative association between *Helicobacter pylori* infection and reflux oesophagitis, Barrett's oesophagus and oesophageal adenocarcinoma, Gut 57 (6) (2008) 721, https://doi.org/10.1136/gut.2007.144774.
[45] K. Sugano, J. Tack, E.J. Kuipers, et al., Kyoto global consensus report on *Helicobacter pylori* gastritis, Gut 64 (9) (2015) 1353–1367, https://doi.org/10.1136/gutjnl-2015-309252.
[46] P. Katelaris, R. Hunt, F. Bazzoli, et al., World Gastroenterology Organisation Global Guidelines: *Helicobacter pylori*, 2021. https://www.worldgastroenterology.org/UserFiles/file/guidelines/helicobacter-pylori-english-2021.pdf. (Last accessed 11 May 2022).
[47] M.A. Akinola, T.A. Oyedele, K.O. Akande, et al., Gastroesophageal reflux disease: prevalence and extraesophageal manifestations among undergraduate students in South

West Nigeria, BMC Gastroenterol. 20 (1) (2020) 160, https://doi.org/10.1186/s12876-020-01292-1.

[48] J. Li, G.I. Perez-Perez, *Helicobacter pylori* the latent human pathogen or an ancestral commensal organism, Front. Microbiol. 9 (2018) 609, https://doi.org/10.3389/fmicb.2018.00609.

[49] H.B. el-Serag, A. Sonnenberg, Opposing time trends of peptic ulcer and reflux disease, Gut 43 (3) (1998) 327–333, https://doi.org/10.1136/gut.43.3.327.

[50] H.B. El-Serag, A. Sonnenberg, M.M. Jamal, J.M. Inadomi, L.A. Crooks, R.M. Feddersen, Corpus gastritis is protective against reflux oesophagitis, Gut 45 (2) (1999) 181–185, https://doi.org/10.1136/gut.45.2.181.

[51] A. Raghunath, A.P.S. Hungin, D. Wooff, S. Childs, Prevalence of *Helicobacter pylori* in patients with gastro-oesophageal reflux disease: systematic review, BMJ (Clinical research ed) 326 (7392) (2003) 737, https://doi.org/10.1136/bmj.326.7392.737.

[52] H. Chiba, T. Gunji, H. Sato, et al., A cross-sectional study on the risk factors for erosive esophagitis in young adults, Intern. Med. 51 (11) (2012) 1293–1299, https://doi.org/10.2169/internalmedicine.51.7241.

[53] T. Gunji, H. Sato, K. Iijima, et al., Risk factors for erosive esophagitis: a cross-sectional study of a large number of Japanese males, J. Gastroenterol. 46 (4) (2011) 448–455, https://doi.org/10.1007/s00535-010-0359-5.

[54] J. Zhang, X.-L. Chen, K.-M. Wang, X.-D. Guo, A.-L. Zuo, J. Gong, Relationship of gastric *Helicobacter pylori* infection to Barrett's esophagus and gastro-esophageal reflux disease in Chinese, World J. Gastroenterol. 10 (5) (2004) 672–675, https://doi.org/10.3748/wjg.v10.i5.672.

[55] H. Nordenstedt, M. Nilsson, R. Johnsen, J. Lagergren, K. Hveem, *Helicobacter pylori* infection and gastroesophageal reflux in a population-based study (The HUNT Study), Helicobacter 12 (1) (2007) 16–22, https://doi.org/10.1111/j.1523-5378.2007.00466.x.

[56] T. Koike, S. Ohara, H. Sekine, et al., *Helicobacter pylori* infection inhibits reflux esophagitis by inducing atrophic gastritis, Am. J. Gastroenterol. 94 (12) (1999) 3468–3472, https://doi.org/10.1111/j.1572-0241.1999.01593.x.

[57] T. Koike, S. Ohara, H. Sekine, et al., Increased gastric acid secretion after *Helicobacter pylori* eradication may be a factor for developing reflux oesophagitis, Aliment. Pharmacol. Ther. 15 (6) (2001) 813–820, https://doi.org/10.1046/j.1365-2036.2001.00988.x.

[58] M. Mihara, Low prevalence of *Helicobacter pylori* infection in patients with reflux esophagitis, Gut 39 (1996) A94.

[59] K. Haruma, H. Hamada, M. Mihara, et al., Negative association between *Helicobacter pylori* infection and reflux esophagitis in older patients: case-control study in Japan, Helicobacter 5 (1) (2000) 24–29, https://doi.org/10.1046/j.1523-5378.2000.00003.x.

[60] B.E. Schenk, E.J. Kuipers, E.C. Klinkenberg-Knol, S.A. Eskes, S.G. Meuwissen, *Helicobacter pylori* and the efficacy of omeprazole therapy for gastroesophageal reflux disease, Am. J. Gastroenterol. 94 (4) (1999) 884–887, https://doi.org/10.1111/j.1572-0241.1999.982_e.x.

[61] A.P. Weston, A.S. Badr, M. Topalovski, R. Cherian, A. Dixon, R.S. Hassanein, Prospective evaluation of the prevalence of gastric *Helicobacter pylori* infection in patients with GERD, Barrett's esophagus, Barrett's dysplasia, and Barrett's adenocarcinoma, Am. J. Gastroenterol. 95 (2) (2000) 387–394, https://doi.org/10.1111/j.1572-0241.2000.01758.x.

[62] P. Bercík, E.F. Verdú, D. Armstrong, et al., The effect of ammonia on omeprazole-induced reduction of gastric acidity in subjects with *Helicobacter pylori* infection, Am. J. Gastroenterol. 95 (4) (2000) 947–955, https://doi.org/10.1111/j.1572-0241.2000.01983.x.

[63] E.J. Kuipers, A.M. Uyterlinde, A.S. Peña, et al., Increase of *Helicobacter pylori*-associated corpus gastritis during acid suppressive therapy: implications for long-term safety, Am. J. Gastroenterol. 90 (9) (1995) 1401–1406.

[64] F. Mazzoleni, L.E. Mazzoleni, C.F. de Magalhães Francesconi, et al., Potential roles of *Helicobacter pylori* treatment, body mass index and waist circumference in the causation of erosive esophagitis: a randomized clinical trial (HEROES-GERD), Int. J. Obes. (London) 44 (1) (2020) 147–158, https://doi.org/10.1038/s41366-019-0391-3.

[65] D. Boltin, Y. Niv, Ghrelin, *Helicobacter pylori* and body mass: is there an association? Isr. Med. Assoc. J. 14 (2) (2012) 130–132.

[66] H. Isomoto, H. Ueno, V.A. Saenko, et al., Impact of *Helicobacter pylori* infection on gastric and plasma ghrelin dynamics in humans, Am. J. Gastroenterol. 100 (8) (2005) 1711–1720, https://doi.org/10.1111/j.1572-0241.2005.41492.x.

[67] J.A. Lane, L.J. Murray, I.M. Harvey, J.L. Donovan, P. Nair, R.F. Harvey, Randomised clinical trial: *Helicobacter pylori* eradication is associated with a significantly increased body mass index in a placebo-controlled study, Aliment. Pharmacol. Ther. 33 (8) (2011) 922–929, https://doi.org/10.1111/j.1365-2036.2011.04610.x.

[68] J. Labenz, A.L. Blum, E. Bayerdörffer, A. Meining, M. Stolte, G. Börsch, Curing *Helicobacter pylori* infection in patients with duodenal ulcer may provoke reflux esophagitis, Gastroenterology 5 (1997) 1442–1447.

[69] G. Holtmann, C. Cain, P. Malfertheiner, Gastric *Helicobacter pylori* infection accelerates healing of reflux esophagitis during treatment with the proton pump inhibitor pantoprazole, Gastroenterology 117 (1) (1999) 11–16, https://doi.org/10.1016/s0016-5085(99)70544-5.

[70] S.Y. Nam, I.J. Choi, K.H. Ryu, B.C. Kim, C.G. Kim, B.H. Nam, Effect of *Helicobacter pylori* infection and its eradication on reflux esophagitis and reflux symptoms, Am. J. Gastroenterol. 105 (10) (2010) 2153–2162, https://doi.org/10.1038/ajg.2010.251.

[71] H. Inoue, I. Imoto, Y. Taguchi, et al., Reflux esophagitis after eradication of *Helicobacter pylori* is associated with the degree of hiatal hernia, Scand. J. Gastroenterol. 39 (11) (2004) 1061–1065, https://doi.org/10.1080/00365520410008006.

[72] J. Tan, Y. Wang, X. Sun, W. Cui, J. Ge, L. Lin, The effect of *Helicobacter pylori* eradication therapy on the development of gastroesophageal reflux disease, Am. J. Med. Sci. 349 (4) (2015) 364–371, https://doi.org/10.1097/maj.0000000000000429.

[73] B. Qian, S. Ma, L. Shang, J. Qian, G. Zhang, Effects of *Helicobacter pylori* eradication on gastroesophageal reflux disease, Helicobacter 16 (4) (2011) 255–265, https://doi.org/10.1111/j.1523-5378.2011.00846.x.

[74] B. Delaney, K. McColl, Review article: *Helicobacter pylori* and gastro-oesophageal reflux disease, Aliment. Pharmacol. Ther. 22 (s1) (2005) 32–40, https://doi.org/10.1111/j.1365-2036.2005.02607.x.

[75] P. Malfertheiner, F. Megraud, C.A. Morain, et al., Management of *Helicobacter pylori* infection—the Maastricht V/Florence Consensus Report, Gut 66 (1) (2017) 6, https://doi.org/10.1136/gutjnl-2016-312288.

[76] W.D. Chey, G.I. Leontiadis, C.W. Howden, S.F. Moss, ACG clinical guideline: treatment of *Helicobacter pylori* infection, Am. J. Gastroenterol. 112 (2) (2017).

[77] H.B. El-Serag, J.Y. Kao, F. Kanwal, et al., Houston consensus conference on testing for *Helicobacter pylori* infection in the United States, Clin. Gastroenterol. Hepatol. 16 (7) (2018) 992–1002.e6, https://doi.org/10.1016/j.cgh.2018.03.013.

CHAPTER 6

Esophageal dysbiosis and neoplasia: Moving from Barrett's esophagus to adenocarcinoma

Andrea Proaño-Vasco[a,b,c] and Michael Quante[a,b]
[a]University Medical Center of Freiburg, Clinic for Internal Medicine II, Gastrointestinal Oncology, Freiburg, Germany
[b]Spemann Graduate School of Biology and Medicine (SGBM), University of Freiburg, Freiburg, Germany
[c]Faculty of Biology, University of Freiburg, Freiburg, Germany

Introduction

Esophageal adenocarcinoma (EAC) is endemic, occurs in industrialized countries, and is a major cause of cancer morbidity and mortality. Despite this, our understanding of EAC pathogenesis and progression's risk to EAC from its only established precursor lesion Barrett's Esophagus (BE) [1–3] is still not fully understood. Although an increasing trend of EAC incidence has been described since 1975 [4], recent data suggest a plateaued incidence since 2010 [5]. Despite diagnostic and therapeutic advances, the prognosis of EAC remains poor [5], highlighting the need for a better understanding of the disease to improve the prognosis and decrease incidence rates. Typically, EAC arises in the distal esophagus near the gastroesophageal junction (GEJ) and is the last stage of a pathological succession of metaplasia, low- and high-grade dysplasia (LGD and HGD) [1,2]. BE is asymptomatic [6] and normally diagnosed by endoscopic screening [7]. It results in an adaptation to a reflux-mediated microenvironment and is characterized by the replacement of squamous epithelium by columnar-like epithelium [7,8], likely through the expansion of gastric/cardia progenitors and their lineage into the esophagus [9]. Multiple genetic alterations have been described during the progression from BE to EAC [10]. These include many somatic mutations such as the inactivation of the tumor suppressor genes TP53 and Cdkn2a, the activation of the oncogenes KRas and Rb [11–15], chromosomal alterations and rearrangements [16,17], and epigenetic modifications [18]. Besides the genetic and pathohistological background of BE

Esophageal Disease and the Role of the Microbiome
https://doi.org/10.1016/B978-0-323-95070-1.00013-8

and EAC, recent data have shown that the esophagus hosts a specific and unique microbiome that changes in relation to the different disease conditions [19]. A strong chronological correlation exists between the decline in *H. pylori* infection rates and the rise in EAC [20]. Yet, the decline of *H. pylori* infection is not an isolated event, but is likely a marker for the dramatic changes that have occurred in the gastrointestinal microbiome, primarily by the widespread of antibiotics as well as increased hygiene conditions or specific proton pump inhibitor (PPI) treatment in westernized nations. Indeed, studies in BE patients have revealed significant changes in the esophageal microbiome [21]. Moreover, unhealthy dietary preferences and overeating that promote obesity, could account for some of the increase in EAC and are likely associated with different colonization of the gut microbiome. Because microbial communities are complex and sensitive ecosystems, they may provide good biomarkers either by responding to relevant changes in the BE epithelium or may even drive progression themselves. The mucinous crypts of BE act as a kind of incubator for microbes adapted to that metaplastic environment. From ecology we know that most organisms gain an advantage from niche-construction and -extension . If there are microbes that are particularly adapted to BE crypts composed of cells with proto-oncogene mutations, then those microbes would be under selective pressure to trigger crypt division so as to enlarge their niche.

If microbes have a causative role in BE progression, they would represent an easily modifiable, and nontoxic avenue toward improved cancer prevention. Thus, better understanding of the tissue-specific and related microbiome (esophagus and gut), as well as the microbial metabolic and inflammatory pathways triggering disease and carcinogenesis might be essential to explain how certain risk factors influence disease progression. In this chapter, we will delve into the link between the microbiome and the carcinogenic progression within the gastroesophageal junction.

Development of microbiome research contributes to a new concept of health and disease

The field of microbiome research started back in the 17th century. Its progression has mainly been driven by the development of new techniques and equipment, and it has led to paradigm shifts in our understanding of health and disease. The relatively recent microbiome research has already delivered novel and important concepts for describing host-microbial interactions such as the metaorganism concept, which considers eukaryotes together with their microbiota as a functional unit [22]. Recently, unique microbial communities have been identified across healthy esophagus, BE, and the different progression stages of EAC. However, there is a lack of studies determining dynamic changes in these communities and their potential causal role in cancer formation. An altered esophageal microbiome has been proposed to contribute to the EAC progression [7]. Establishing causative links between microbiota and the different pathological stages of the distal esophagus is still in its infancy [7]. To date, no study has yet shown if the microenvironment of the different EAC-pathological stages attracts certain types of microbes [23]. However, localizing microbes unique to the different pathological EAC-stages and or microbes influencing the EAC-progression could lead to microbe-specific biomarkers or alternative microbe-targeted therapies. Considering that each person's individualized and highly distinct microbial community might affect EAC progression, microbe-targeted therapies could also be personalized [7].

Risk and environmental factors altering the risk of EAC-progression

Risk factors of EAC may be linked to the modern lifestyle of western, industrialized civilizations. These include tobacco and alcohol consumption, NSAID intake [24–26], as well as other risk factors strongly associated with obesity such as a diet high in processed food, saturated fat, refined sugars and salt, while low in polyunsaturated fats, fiber, and micronutrients [26–29]. In the past, it has been proposed, that those obesity-linked alterations of the microbiome increase inflammation in the gastrointestinal tract (GIT), potentially providing an environment that facilitates the development of EAC [7,30,31]. Gastrointestinal microbiota contributes to digestion, energy recovery, and metabolism of food components that are partly indigestible to the host [32,33]. Thus, the composition of microbial communities, as well as their functional potential is greatly shaped by the diet composition of the host [19]. Furthermore, a microbiota shaped by an unbalanced diet can contribute to the development of metabolic disorders, local immune cell activation, systemic inflammatory effects [33,34], and ultimately, to the development of certain cancers. For instance, in a genetic mouse model of BE, L2-IL1B, we earlier showed that a high-fat diet (HFD) and a high fructose diet (HFrD) alter the composition of intestinal microbial communities while affecting the development of EAC [35,36]. Although the effect of the HFrD was less pronounced than the effect of the HFD, a diet-related shift of the intestinal microbiota was observed in both cases. This shift was associated with disease progression and suggested that the intestinal microbiome indirectly affects tissues distant to the gut systemically through the absorption of microbial metabolites.

Same applies to PPIs, which are commonly used to relieve symptoms, heal, and prevent different gastroesophageal pathologies such as gastroesophageal reflux disease (GERD), esophagitis, or peptic ulcer disease (PUD). The use of PPIs has been associated with reduced risk of BE-progression and thus, most patients with BE are prescribed PPIs [7,37]. The potential effects that PPIs might have on the esophageal microbiome have been discussed earlier [38,39]. An active intake of PPIs has been linked to a reduced relative abundance of gram-negative bacteria and an increased relative abundance of *Streptococcus* spp. [38]. Considering that gram-negative bacteria are thought to promote chronic inflammation [22,40,41], the achieved decrease of gram-negative bacteria by prescribing patients with PPIs might help to prevent EAC. While the prophylactic role of PPIs on EAC carcinogenesis is still under discussion, PPIs have also been associated with alterations in the gut microbiome [42], which might indirectly also have an impact on the pathogenesis of EAC, similar to a change in diet.

Microbial alterations in the esophagus linked to histopathological changes

Different microbes can thrive in slightly different microenvironments across the gastrointestinal tract with varying oxygen- and pH-levels [7], as well as local mucus layer and present antimicrobial proteins [43,44]. Similar to other luminal tissues of the gastrointestinal tract (GIT), the esophagus provides a potential environment for bacterial colonization [45]. The majority of the overall microbiome consists of bacteria located in the colon. The esophageal microbiome makes up only a small part of the whole microbiome and little is known about the presence and nature of bacteria colonizing the esophagus [46]. Aiming to assess the potential role of the esophageal microbiome on BE and EAC, as well as the link of the esophageal microbiome to their risk factors, the following observations have been made in the past.

Initial studies characterized the specific esophageal microbiome with advanced sequencing approaches for that time relay in 2004 (16S rDNA PCR). Members of Firmicutes, Bacteroidetes, Actinobacteria, Proteobacteria, Fusobacteria, and TM7 were detected in healthy human esophageal tissue, and additionally, the absence of bacteria typically colonizing the oral cavity provided evidence for a specific esophageal microbiota [45]. Further studies confirmed these initial findings and described gram-positive bacteria to be dominant in the esophagus [47,48]. Other studies characterizing microbial communities during EAC-progression have shown some trends and microbial alterations in the esophagus to be related to pathological conditions but also to age [49]. While the microbiome of a healthy esophagus has been characterized by gram-positive bacteria [50], in the microbiome of esophageal tissue from GERD- and BE-patients, an increased abundance of gram-negative bacteria was found [47,50,51]. While *Streptococcus*—a member of the phyla Firmicutes, was shown to be the most abundant bacteria in esophageal tissue of healthy, non-BE patients, *Veillonella*—also a member of the phyla Firmicutes, was shown to be the most abundant bacteria in esophageal tissue of BE patients [52]. Additionally, while the relative abundance of *Streptococcus* was shown to decrease in esophageal tissue of BE patients [53], *Veillonella* was not found in healthy tissue [52,53].

To date, there are some characterizing microbial communities in EAC-patients. However, the data are limited and further studies are required [7]. Current studies assessing and characterizing microbial communities across the pathological progression from BE to LGD, HGD, and EAC show a substantial shift in the esophageal microbiome between the stages LGD and HGD [38] and support the potential role of the microbiome in the EAC-progression [7]. In comparison to BE- or LGD-patients, HGD- or EAC-patients showed a significant increase in the phylum Proteobacteria and a significant decrease in the phylum Firmicutes. Among the Proteobacteria, the family *Enterobacteriaceae* was increased in HGD/EAC-patients compared to BE/LGD-patients [38]. Additionally, the species *Akkermansia muciniphila* of the phyla Verrucomicrobia was found in HGD/EAC-patients and not in BE/LGD-patients [38,54]. Interestingly, *Enterobacteriaceae* is thought to induce inflammation in the gut of patients with inflammatory bowel disease (IBD) and has been proposed to have a similar effect in the esophagus of EAC-patients [38,55], *A. muciniphila* has been generally associated with increased tissue inflammation [38,54]. Moreover, EAC-patients show a decreased microbial alpha-diversity compared to BE-, LGD-, and HGD-patients (Simpson and Shannon index) [38,56]. The evaluation of beta-diversity using the Bray–Curtis measure showed significant differences between microbial populations of EAC- and healthy patients and between EAC- and BE-patients [53,56].

On the whole, studies have shown an enrichment of gram-positive bacteria in the esophagus of healthy individuals and an enrichment of gram-negative bacteria in patients with metaplastic esophagus. Of note, bacteria found in the esophagus of HGD- and EAC-patients were associated with inflammation. Therefore, this shift from gram-positive to gram-negative bacteria appears to be a further indicator of the beginning of pathogenesis. Fig. 1 shows the histopathological EAC-progression in relation to some risk factors and the microbial composition of the esophagus.

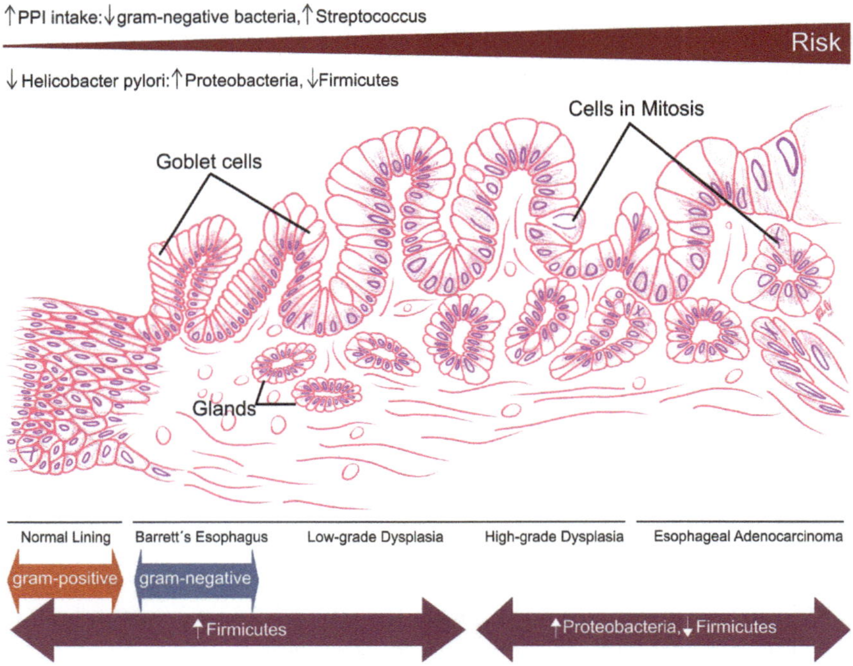

Fig. 1 Pathological stages of EAC in relation the bacterial communities identified across healthy esophagus, BE, and the progression stages of EAC. The cartoon of the histopathological EAC-progression was adapted by Paula Proaño Vasco from the figure provided by the Pathology Department of the Johns Hopkins University (https://pathology.jhu.edu/barretts-esophagus/dysplasia). An increase in PPI intake and a reduction of *H. pylori* (gram-negative) inversely correlate with an increased risk of EAC-carcinogenesis. The microbiota of a healthy esophagus has been characterized by gram-positive bacteria and gram-negative bacteria is abundant in the esophageal tissue of BE-patients. In addition, a PPI intake is associated with alterations of intestinal communities and has been linked to a reduced abundance of gram-negative bacteria and an increased abundance of *Streptococcus* (Firmicutes). While in healthy, BE- and LGD-tissue Firmicutes are dominant, in HGD- and EAC-tissue, there is a low abundance of Firmicutes and a high abundance of Proteobacteria.

Microbes influencing BE- and EAC-pathogenesis

Although microbial footprints linked to the different pathological stages are evident, it is not clear whether carcinogenesis results in part due to the microbiome itself or whether the different microbial footprints result in response to changes in the tissue environment caused by carcinogenesis. Hence, the microbial influence in the premalignancy and malignancy of EAC is still unclear.

Microbes live on and within macroorganisms such as animals and plants and form an integrated entity, the metaorganism [57]. Microbes also undergo Darwinian evolution and have both beneficial and harmful effects on other organisms [58,59]. The way microorganisms survive, evolve, and thrive within the human body is still not well-understood. However, mathematical models of evolution demonstrate that alterations in microbial communities present in different environments arise as a result of neutral evolution [57]. Furthermore, to better understand the evolution of microbial populations, it is necessary to consider that microbes can migrate into and within the body. This can lead to inconstancy in the local microbial diversity and then be reflected in the health status of the host [7]. For instance, a decreased microbial diversity in the intestine is regarded as a sign of an unbalanced, "unhealthy" microbiome linked to an increased risk of developing obesity and related diseases [60]. The microbial communities of the esophagus are different than the microbial communities of the stomach and oral cavity [7,19]. It is still not clear whether the acidic environment prevalent in the BE represents a selective advantage for certain microbes over others [7]. Nonetheless, a better insight into the microbial migration from other parts of the gastrointestinal tract to the distal esophagus might also help to understand the different bacterial compositions along EAC-progression.

The most prominent example is *H. pylori*, a gram-negative bacterium within the Proteobacteria phylum [61]. While *H. pylori* is positively associated with gastroduodenal PUD and noncardia gastric cancer, it is inversely associated with adenocarcinoma of the gastric cardia [61] and EAC [7,51,62]. These trends match with the epidemiological observation that while the incidence of noncardia gastric cancer is declining, the incidence of cardia gastric cancer is increasing [63]. Interestingly, these trends are related to the "westernization" of nations, the increasing rates of obesity and metabolic syndrome [7], and the decrease of *H. pylori* across the population as a result of the widespread use of antibiotics [6,64]. Furthermore, the implication that carcinogenesis profits from a reduction of *H. pylori* [51] is based on the suggestion that it plays an important role in the negative selection of cells that undergo chromosomal alterations. In addition, the absence of *H. pylori* in the gastric microbiome results in the increase of the phylum Proteobacteria and the decrease of the phylum Firmicutes [6]. The causal role of *H. pylori* in EAC-carcinogenesis is still discussed and implies the need for more mechanistic studies in this area [7].

Esophageal microbiome: current techniques of sampling and analysis need to be improved

To assess the role of microbes on BE- and EAC-pathogenesis, it is important to first understand the potential influence of microbes on the selection of certain mutated clones or cancer-primed genotypes. Since the joint analysis of host and microbial genomes in the same samples is not common and easily available yet, questions addressing the potential of inter- and/or intracellular microbes to increase mutagenesis or promote tumor spread remain unanswered. Further, the potential of microbes to be used as spatial or prediction markers in patients with premalignant conditions remains unknown. So far, most studies assessing the microbiome-cancer relationship are association studies. Thus, before concluding microbes induce cancer growth or metastasis, more experimental models are required in order to reveal the differences between the association and causation of microbes in cancer [7]. In case of EAC, these association studies are mostly performed on samples taken from patients of different pathological stages during endoscopic brushings or forceps biopsies [65]. The microbial analysis of those samples is then used to draw associations between the microbes and the disease stages [7]. More controlled human trials focusing on causative mechanisms are therefore needed. However, the conduction of those studies is particularly difficult owing to the many technical challenges that frequent microbial sampling from the esophagus represents [56,66,67]. Most sampling methods of esophageal tissue are invasive for the patient. Hence, the collection of microbial samples is expensive and not efficient enough, representing an issue for data generation and analysis. Host contamination, microbial variation dependent on the time of day, as well as inconsistency in sampling during fasting and postprandial periods are other factors that have to be considered and might alter the accuracy of the results and conclusions. However, a decreased esophageal specificity remains one of the main challenges to overcome [7]. Furthermore, a better understanding of the role of microbiota on

BE- and EAC-pathogenesis implies the increase of datasets available for esophageal microbiome. To achieve this, easier but more efficient sampling methods are needed. Even with their drawbacks such as low esophageal specificity, the Cytosponge device, saliva, and oral swabs are alternative, noninvasive sampling techniques for detecting microbial changes in BE or EAC. The Cytosponge device can measure the majority of genera found in endoscopic brushings or forceps biopsies [56]. When comparing microbial communities of saliva from healthy individuals to patients with reflux esophagitis, differences in oral microbial taxa were made [68]. Microbes from oral swabs have also been shown to highly correlate with the esophageal microbiome [69]. Although more studies regarding the accuracy of these alternative methods are needed, they seem to represent fast, complementary diagnostic tools should be improved in the future.

Random mutations arise as result of DNA replication errors and many cancers accrue in single cells through (*epi*)genetic changes that induce a malignant phenotype [7]. When assessing the applicability of the microbiome in cancer diagnostics [70,71], the corresponding microbial DNA in 32 different cancer types within The Cancer Genome Atlas (TCGA) was analyzed and microbial variations between healthy and cancerous samples were found. The application of machine learning algorithms to whole genome sequencing (WGS) data has allowed a tissue-specific microbiome classification and designation between stages I and IV of different types of cancer such as colon adenocarcinoma, stomach adenocarcinoma, and kidney renal clear cell carcinoma [70]. Microbiome research has been mainly driven by samples coming from the colon due to an easy and noninvasively collection method. In addition, the bacterial content of these samples is much greater than the bacterial content of samples coming from other organs such as the esophagus. Hence, it remains a challenge to determine the benefit and applicability of microbiome diagnostics for the cancer of organs where the microbial communities are less abundant [72].

There is a need to perform more standardized, longitudinal studies in humans to better track microbial changes in the tissue at different disease stages. Additionally, methodologic processes are needed to track possible microbial migrations within the body in response to certain cancer-related changes in the microenvironment.

After resolving the different sampling issues described earlier, there is a need to create and support open access datasets and improve the accuracy of the microbiome's characterization and classification in order to reveal more accurate associations and causalities between microbes and the host's health.

An example of a public accessible microbial dataset is the American Gut Project [7], which includes data from fecal, oral, and/or skin samples from over 10,000 individuals from 57 countries [42]. Most of the current datasets available are based on data obtained from 16S rRNA-sequencing analysis. In contrast, data obtained from WGS analysis can be used to address microbial research questions and additionally to characterize the host genome [7]. Another example of an initiative aiming to better understand the human microbiome is the NIH Integrative Human Microbiome Project (iHMP). Although it focuses on diseases other than cancer, the iHMP contains longitudinal datasets from microbiomes and host cell genomes and integrates functional data from metagenomics and -transcriptomics [73]. To date, the esophageal microbiome has mainly been assessed by 16S-rRNA sequencing. Considering that the extraction yield of bacteria contained in esophageal samples is much lower than of, e.g., intestinal samples, alternative technologies such as WGS have been proposed to yield more accurate results [7]. Despite considerably higher costs compared to 16S-rRNA sequencing, WGS yields a higher number of genera and shows a significantly higher alpha diversity in soil samples [74]. Hence, the application of the WGS technology in human cancer studies represents a research possibility and step forward in the microbial research of cancer and of the EAC pathology.

Conclusion

Carcinogenesis is characterized by alterations in cell genotypes that lead to advantageous cell phenotypes and is affected by different (*epi*)genetic and environmental factors [7]. Understanding microbial changes within precancerous and cancerous cells remains a challenge and defining the host's microbiome within precancer and cancer is thus, a research field of great importance, as it is for BE and EAC. Based on microbial profiles, findings in this field would be beneficial for qualifying the overall risk of BE-development and the risk for BE-patients to develop EAC, improving BE surveillance and early and late EAC detection and treatment.

The multiple links between the microbiome and EAC-development made so far, suggest that alterations in the microbial communities might contribute to inflammatory or carcinogenic changes in the microenvironment of the lower esophagus by interfering with the immune system and metabolism of the host, and the defense against pathogens [19]. Many of these links remain correlative owing to the difficulty of modeling host-microbe relationships in a meaningful way [75] and fail to prove direct functional causalities. More data are needed to draw conclusions on this matter.

Future directions

- Defining the relation between esophageal microbiome dysbiosis and BE and EAC with more comprehensive and systemic studies to further examine the complex interrelationships between epidemiological, environmental, and genetic factors.
- Identification of best methodologic processes to assess these esophageal disease-related microbial alterations will be important.
- In addtion to the microbial environment in the esophagus, the systemic effect of the bacteria located in other parts of the gastrointestinal tract should also be assessed as it might have an effect on the risks for BE and EAC.
- Development and evaluation of possible mitigation strategies for esophageal dysbiosis associated with BE and EAC are needed.

COI disclosures

The authors declare no conflict of interest.

References

[1] K. Curtius, et al., Computational modelling suggests that Barrett's oesophagus may be the precursor of all oesophageal adenocarcinomas, Gut 70 (8) (2020) 1435–1440.
[2] K. Nowicki-Osuch, et al., Molecular phenotyping reveals the identity of Barrett's esophagus and its malignant transition, Science 373 (6556) (2021) 760–767.
[3] A.P. Thrift, Barrett's esophagus and esophageal adenocarcinoma: how common are they really? Dig. Dis. Sci. 63 (8) (2018) 1988–1996.
[4] C.Y. Kong, et al., Exploring the recent trend in esophageal adenocarcinoma incidence and mortality using comparative simulation modeling, Cancer Epidemiol. Biomarkers Prev. 23 (6) (2014) 997.
[5] NIH (National Cancer Institute; Surveillance, Epidemiology, and End Results Program), Cancer Stat Facts: Esophageal Cancer, National Cancer Institute, 2021. https://seer.cancer.gov/statfacts/html/esoph.html.
[6] E.J. Snider, D.E. Freedberg, J.A. Abrams, Potential role of the microbiome in Barrett's esophagus and esophageal adenocarcinoma, Dig. Dis. Sci. 61 (8) (2016) 2217–2225.
[7] C. Guccione, et al., Challenges in determining the role of microbiome evolution in Barrett's esophagus and progression to esophageal adenocarcinoma, Microorganisms 9 (10) (2021), 2003, https://doi.org/10.3390/microorganisms9102003.
[8] J.L. Herrera Elizondo, et al., Prevalence of Barrett's esophagus: an observational study from a gastroenterology clinic, Rev. Gastroenterol. Mex. 82 (4) (2017) 296–300.
[9] M. Quante, T.A. Graham, M. Jansen, Insights into the pathophysiology of esophageal adenocarcinoma, Gastroenterology 154 (2) (2018) 406–420.
[10] M.D. Stachler, et al., Detection of mutations in Barrett's esophagus before progression to high-grade dysplasia or adenocarcinoma, Gastroenterology 155 (1) (2018) 156–167.

[11] C.S. Ross-Innes, et al., Whole-genome sequencing provides new insights into the clonal architecture of Barrett's esophagus and esophageal adenocarcinoma, Nat. Genet. 47 (9) (2015) 1038–1046.
[12] M.D. Stachler, et al., Paired exome analysis of Barrett's esophagus and adenocarcinoma, Nat. Genet. 47 (9) (2015) 1047–1055.
[13] W.M. Grady, M. Yu, Molecular evolution of metaplasia to adenocarcinoma in the esophagus, Dig. Dis. Sci. 63 (8) (2018) 2059–2069.
[14] C. Zhou, J. Li, Q. Li, CDKN2A methylation in esophageal cancer: a meta-analysis, Oncotarget 8 (30) (2017) 50071–50083.
[15] A.M. Frankell, et al., The landscape of selection in 551 esophageal adenocarcinomas defines genomic biomarkers for the clinic, Nat. Genet. 51 (3) (2019) 506–516.
[16] X. Li, et al., Temporal and spatial evolution of somatic chromosomal alterations: a case-cohort study of Barrett's esophagus, Cancer Prev. Res. 7 (1) (2014) 114.
[17] S. Killcoyne, et al., Genomic copy number predicts esophageal cancer years before transformation, Nat. Med. 26 (2020) 1726–1732, https://doi.org/10.1038/s41591-020-1033-y.
[18] M. Yu, et al., Subtypes of Barrett's oesophagus and oesophageal adenocarcinoma based on genome-wide methylation analysis, Gut 68 (3) (2019) 389–399.
[19] T. Baumeister, M. Quante, J. Bornschein, Gut microbiota and oesophageal disease—an update, Microb. Health Dis. 1 (2019), e207.
[20] L. Fischbach, et al., The association between Barrett's esophagus and *Helicobacter pylori* infection: a meta-analysis, Helicobacter 17 (2012) 163–175.
[21] L. Yang, F. Francois, Z. Pei, Molecular pathways: pathogenesis and clinical implications of microbiome alteration in esophagitis and Barrett esophagus, Clin. Cancer Res. 18 (8) (2012) 2138–2144.
[22] G. Berg, et al., Microbiome definition re-visited: old concepts and new challenges, Microbiome 8 (1) (2020) 103.
[23] A. Hasan, et al., Microbiome of the aerodigestive tract in health and esophageal disease, Dig. Dis. Sci. 66 (1) (2021) 12–18.
[24] S. Lipka, A. Kumar, J.E. Richter, Impact of diagnostic delay and other risk factors on eosinophilic esophagitis phenotype and esophageal diameter, J. Clin. Gastroenterol. 50 (2) (2016) 134–140.
[25] L.H. Eusebi, et al., Global prevalence of, and risk factors for, gastro-oesophageal reflux symptoms: a meta-analysis, Gut 67 (3) (2018) 430–440.
[26] F.L. Huang, S.J. Yu, Esophageal cancer: risk factors, genetic association, and treatment, Asian J. Surg. 41 (3) (2018) 210–215.
[27] A. Kubo, et al., Dietary patterns and the risk of Barrett's esophagus, Am. J. Epidemiol. 167 (7) (2008) 839–846.
[28] S.A. Navarro Silvera, et al., Principal component analysis of dietary and lifestyle patterns in relation to risk of subtypes of esophageal and gastric cancer, Ann. Epidemiol. 21 (7) (2011) 543–550.
[29] M.H. Collins, K. Capocelli, G.-Y. Yang, Eosinophilic gastrointestinal disorders pathology, Front. Med. 4 (2018) 261.
[30] J.O. Alemán, et al., Mechanisms of obesity-induced gastrointestinal neoplasia, Gastroenterology 146 (2) (2014) 357–373.
[31] J. Lv, et al., Alteration of the esophageal microbiota in Barrett's esophagus and esophageal adenocarcinoma, World J. Gastroenterol. 25 (18) (2019) 2149–2161.
[32] M. Nieuwdorp, et al., Role of the microbiome in energy regulation and metabolism, Gastroenterology 146 (6) (2014) 1525–1533.
[33] J.L. Sonnenburg, F. Bäckhed, Diet-microbiota interactions as moderators of human metabolism, Nature 535 (7610) (2016) 56–64.

[34] P.J. Turnbaugh, et al., An obesity-associated gut microbiome with increased capacity for energy harvest, Nature 444 (7122) (2006) 1027–1031.
[35] N.S. Münch, et al., High-fat diet accelerates carcinogenesis in a mouse model of Barrett's esophagus via Interleukin 8 and alterations to the gut microbiome, Gastroenterology 157 (2) (2019) 492–506.e2.
[36] A. Proaño-Vasco, et al., High-fructose diet alters intestinal microbial profile and correlates with early tumorigenesis in a mouse model of Barrett's esophagus, Microorganisms 9 (12) (2021) 2432.
[37] L. Dall'Olmo, et al., Role of proton pump inhibitor on esophageal carcinogenesis and pancreatic acinar cell metaplasia development: an experimental in vivo study, PLoS One 9 (11) (2014), e112862.
[38] E.J. Snider, et al., Alterations to the esophageal microbiome associated with progression from Barrett's esophagus to esophageal adenocarcinoma, Cancer Epidemiol. Biomarkers Prev. 28 (10) (2019) 1687.
[39] K.B. Dunbar, R.F. Souza, S.J. Spechler, The effect of proton pump inhibitors on Barrett's esophagus, Gastroenterol. Clin. North Am. 44 (2) (2015) 415–424.
[40] S. Zhou, et al., Gram negative bacterial inflammation ameliorated by the plasma protein beta 2-glycoprotein I, Sci. Rep. 6 (2016) 33656.
[41] J. Wang, W.-D. Chen, Y.-D. Wang, The relationship between gut microbiota and inflammatory diseases: the role of macrophages, Front. Microbiol. 11 (2020) 1065, https://doi.org/10.3389/fmicb.2020.01065.
[42] D. McDonald, et al., American Gut: an open platform for citizen science microbiome research, mSystems 3 (3) (2018). e00031-18.
[43] J.K. Goodrich, et al., Human genetics shape the gut microbiome, Cell 159 (4) (2014) 789–799.
[44] M. Wlodarska, A.D. Kostic, R.J. Xavier, An integrative view of microbiome-host interactions in inflammatory bowel diseases, Cell Host Microbe 17 (5) (2015) 577–591.
[45] Z. Pei, et al., Bacterial biota in the human distal esophagus, Proc. Natl. Acad. Sci. U. S. A. 101 (12) (2004) 4250–4255.
[46] B.E. Lacy, B. Spiegel, Introduction to the gut microbiome special issue, Am. J. Gastroenterol. 114 (7) (2019) 1013.
[47] L. Yang, F. Francois, Z. Pei, Molecular pathways: pathogenesis and clinical implications of microbiome alteration in esophagitis and Barrett esophagus, Clin. Cancer Res. 18 (8) (2012) 2138–2144.
[48] S.A. Fillon, et al., Novel device to sample the esophageal microbiome—the esophageal string test, PLoS One 7 (9) (2012), e42938.
[49] N.P. Deshpande, et al., Signatures within the esophageal microbiome are associated with host genetics, age, and disease, Microbiome 6 (1) (2018) 227.
[50] L. Yang, et al., Inflammation and intestinal metaplasia of the distal esophagus are associated with alterations in the microbiome, Gastroenterology 137 (2) (2009) 588–597.
[51] K.L. Blackett, et al., Oesophageal bacterial biofilm changes in gastro-oesophageal reflux disease, Barrett's and oesophageal carcinoma: association or causality? Aliment. Pharmacol. Ther. 37 (11) (2013) 1084–1092.
[52] N. Liu, et al., Characterization of bacterial biota in the distal esophagus of Japanese patients with reflux esophagitis and Barrett's esophagus, BMC Infect. Dis. 13 (1) (2013) 130.
[53] L.R. Lopetuso, et al., Esophageal microbiome signature in patients with Barrett's esophagus and esophageal adenocarcinoma, PLoS One 15 (5) (2020), e0231789.
[54] M. Derrien, C. Belzer, W.M. de Vos, *Akkermansia muciniphila* and its role in regulating host functions, Microb. Pathog. 106 (2017) 171–181.

[55] W.S. Garrett, et al., Enterobacteriaceae act in concert with the gut microbiota to induce spontaneous and maternally transmitted colitis, Cell Host Microbe 8 (3) (2010) 292–300.
[56] D.R.F. Elliott, et al., A non-endoscopic device to sample the oesophageal microbiota: a case-control study, Lancet Gastroenterol. Hepatol. 2 (1) (2017) 32–42.
[57] M. Sieber, et al., Neutrality in the metaorganism, PLoS Biol. 17 (6) (2019), e3000298.
[58] E.R. Davenport, et al., The human microbiome in evolution, BMC Biol. 15 (1) (2017) 127.
[59] E. Rosenberg, G. Sharon, I. Zilber-Rosenberg, The hologenome theory of evolution contains Lamarckian aspects within a Darwinian framework, Environ. Microbiol. 11 (12) (2009) 2959–2962.
[60] F. Deng, Y. Li, J. Zhao, The gut microbiome of healthy long-living people, Aging 11 (2) (2019) 289–290.
[61] J.G. Kusters, A.H. van Vliet, E.J. Kuipers, Pathogenesis of *Helicobacter pylori* infection, Clin. Microbiol. Rev. 19 (3) (2006) 449–490.
[62] S. Anandasabapathy, et al., Clinical and endoscopic factors predict higher pathologic grades of Barrett dysplasia, Cancer 109 (4) (2007) 668–674.
[63] S.C. Shah, et al., Population-based analysis of differences in gastric cancer incidence among races and ethnicities in individuals age 50 years and older, Gastroenterology 159 (5) (2020) 1705–1714.e2.
[64] A.P. Thrift, Global burden and epidemiology of Barrett oesophagus and oesophageal cancer, Nat. Rev. Gastroenterol. Hepatol. 18 (6) (2021) 432–443.
[65] M. May, J.A. Abrams, Emerging insights into the esophageal microbiome, Curr. Treat Options Gastroenterol. 16 (1) (2018) 72–85.
[66] M. Freeman, et al., Acceptability of the cytosponge procedure for detecting Barrett's oesophagus: a qualitative study, BMJ Open 7 (3) (2017), e013901.
[67] V. Sanghi, P.N. Thota, Barrett's esophagus: novel strategies for screening and surveillance, Ther. Adv. Chronic Dis. 10 (2019), 2040622319837851.
[68] B. Wang, et al., Patients with reflux esophagitis possess a possible different oral microbiota compared with healthy controls, Front. Pharmacol. 11 (2020) 1000, https://doi.org/10.3389/fphar.2020.01000.
[69] M.K. Annavajhala, et al., Relationship of the esophageal microbiome and tissue gene expression and links to the oral microbiome: a randomized clinical trial, Clin. Transl. Gastroenterol. 11 (12) (2020), e00235.
[70] G.D. Poore, et al., Microbiome analyses of blood and tissues suggest cancer diagnostic approach, Nature 579 (7800) (2020) 567–574.
[71] S.S. Manzoor, A. Doedens, M.B. Burns, The promise and challenge of cancer microbiome research, Genome Biol. 21 (1) (2020) 131.
[72] R. Sender, S. Fuchs, R. Milo, Revised estimates for the number of human and bacteria cells in the body, PLoS Biol. 14 (8) (2016), e1002533.
[73] L.M. Proctor, et al., The integrative human microbiome project, Nature 569 (7758) (2019) 641–648.
[74] K.D. Brumfield, et al., Microbial resolution of whole genome shotgun and 16S amplicon metagenomic sequencing using publicly available NEON data, PLoS One 15 (2) (2020), e0228899.
[75] J. Puschhof, et al., Intestinal organoid cocultures with microbes, Nat. Protoc. 16 (10) (2021) 4633–4649.

CHAPTER 7

Esophageal dysbiosis and esophageal squamous cell carcinoma

Alyyah Malick, Ali Soroush, and Julian A. Abrams
Division of Digestive and Liver Diseases, Department of Medicine, Columbia University Irving Medical Center, New York, NY, United States

Introduction

Esophageal cancer is the 10th most common incident cancer globally, with approximately 604,000 new cases in 2020, and the sixth most common cause of cancer deaths worldwide [1]. Typically, esophageal cancer is associated with a poor prognosis as symptoms appear late and patients are thereby diagnosed at advanced stages. In the United States, median survival is 10 months and 5 year survival rate is 22% [2]. The most common subtype of esophageal cancer globally is esophageal squamous cell carcinoma (ESCC), accounting for 84% of cases. ESCC poses a major public health burden, especially in high incidence areas such as South America, eastern and central Asia (the "Asian esophageal cancer belt") and southeastern Africa (the "African esophageal cancer corridor"). In the United States, ESCC disproportionately affects African- and Asian-American patients [3]. Over the past decades, ESCC rates have been declining or stabilizing globally, first in high-income countries and now in middle-income countries [4–8]. The geographic and temporal variation in ESCC incidence implies that modifiable environmental risk factors play a large role in ESCC carcinogenesis.

The most studied risk factors for ESCC are tobacco smoking and alcohol consumption, which result in exposure to known carcinogens [9]. Consumed ethanol is metabolized by alcohol dehydrogenase to acetaldehyde, which forms DNA adducts, inhibits DNA damage responses, and increases production of reactive oxygen species [10]. Tobacco smoking results in exposure to *N*-nitrosamines and polycyclic aromatic hydrocarbons that form DNA adducts and synergistically promote cancer development along with acetaldehyde [11,12]. In Western countries, ESCC is primarily associated with smoking and alcohol consumption, and these exposures may to some

Esophageal Disease and the Role of the Microbiome
https://doi.org/10.1016/B978-0-323-95070-1.00014-X

extent account for increased risk based on race and gender [13–15]. In non-Western countries with high ESCC incidence, alcohol and smoking may play a less important role in ESCC etiology, and other risk factors have been identified [9,16,17]. These include intake of hot food and beverages [18,19], pickled vegetables [20], low fruit and vegetable consumption [19,21,22], micronutrient deficiencies [23], opium [24,25] and betel quid usage [26–28], and exposure to household pollutants such as unpiped water [16,19,29] and certain fuels [19,30]. Low socioeconomic status is one of the most consistent risk factors for ESCC across all regions, even after adjustment for factors such as age, alcohol, and tobacco. Other exposures mentioned above may partially explain this association [17,31,32].

Poor oral health is another well-established risk factor for ESCC, leading to hypotheses that the oral microbiome, and by extension, the esophageal and gastric microbiome, may be involved in ESCC carcinogenesis [33]. Interventions on the oral microbiome have been shown to impact the esophageal microbiome and esophageal gene expression, likely modifying disease risk [34]. There is increasing evidence that bacterial microbiota play a role in mediating the effects of known environmental risk factors or act through other pathways with respect to carcinogenesis, tumor progression, and response to treatment [35]. Recent studies have capitalized on methodological advances to characterize oral, esophageal, and gastric microbiome alterations in individuals with ESCC and explore whether certain bacteria may play a key role in cancer development and progression.

Esophageal dysbiosis in Esophageal squamous cell carcinoma

Most studies of the esophageal microbiome have found reduced alpha diversity (hereafter referred to as "diversity"), which represents lower microbial richness and diversity, in ESCC patients compared with controls [36–38] or in ESCC tumor tissue compared with nontumor tissue [39]. Diversity has been found to decrease as the disease progresses from precursor lesions to carcinoma, with higher diversity in low-grade dysplasia compared with high-grade dysplasia and ESCC [40]. However, diversity is not significantly different across different ESCC tumor stages or based on lymph node metastasis [41,42]. Reduction in diversity persists in ESCC patients 9–12 months postradical esophagectomy [37]. Changes in diversity may be affected by factors such as exposure history and tumor histology. A cross-sectional study in Russia found that changes in diversity depended on tumor stroma type;

ESCC tumors with high CD8+ T cell counts had significant decreases in diversity compared with tumor tissue with low CD8+ T cell counts [43]. Alcohol use is also associated with changes in diversity, with ESCC non-drinkers having increased tumor-associated microbiome diversity compared with ESCC drinkers [44].

Studies of the esophageal microbiome have generally found differences in global microbiome composition as assessed by beta diversity between ESCC and healthy controls or nontumor tissue. There are various reports of differentially abundant taxa, with mixed and occasionally contradictory findings, which are summarized in Table 1. Multiple studies have found an increase in relative abundance of the phyla *Fusobacterium* and *Bacteroidetes* with ESCC and decrease in relative abundance of *Firmicutes* [36,39,42,47]. Several studies of the esophageal microbiome in ESCC have observed increases in the relative abundance of the genus *Streptococcus* with progressive tumor stage [39–41]. Relative abundance of *Streptococcus*, when analyzed in conjunction with *Neisseria* or *Prevotella* abundance, has been found to be associated with ESCC and ESCC progression and survival [41,47]. Further study is required to assess species-specific association and temporal trends of *Streptococcus* in the esophageal microbiome as well as possible mechanisms of cancer promotion.

Although the findings from esophageal microbiome studies have had varied results with regard to taxonomic alterations, two oral pathogens have consistently been found to be associated with ESCC: *Porphyromonas gingivalis* and *Fusobacterium nucleatum* (Fig. 1). In the following sections, we provide a more in-depth discussion of these two bacterial species.

P. gingivalis

P. gingivalis is increased in relative abundance in the microbiome of ESCC, and several studies have focused on elucidating its potential role in tumor progression. *P. gingivalis* is a known etiological factor in periodontal disease, which has been associated with esophageal cancer [48,49]. *P. gingivalis* sequesters inside phagocytic cells, impairs host defenses, and causes shifts in the gingival microbial community along with inflammation that promotes periodontal bone destruction [50]. Given the proximity of the oral cavity and esophagus, as well as the overlap in microbiomes and histologic similarity of oral and esophageal squamous cells, it has been postulated that *P. gingivalis* may also play a role in ESCC.

Table 1 Summary of studies characterizing esophageal microbiome in ESCC.

Author (year)	Study Design	Country	Cases (N)	Controls (N)	Sample Type	Method	Alpha Diversity	Differentially Abundant Taxa
Yang et al. (2021) [36] PMID: 33681225	Case–control	China, Guangzhou (Guangdong province)	18 ESCC	11 healthy controls	Esophageal biopsy	16S rRNA, NGS	Lower in ESCC than controls	Relative abundance of *Fusobacterium*, *Bacteroidetes*, and *Spirochaetes* increased and relative abundance of *Proteobacteria* and *Thermi* decreased in ESCC.
Li, Zhengqi et al. (2021) [40] PMID: 34604107	Cross-sectional	China, Beijing, Henan province, Hebei province, Shandong province	60 Low-grade dysplasia (LGD), 64 high-grade dysplasia (HGD), 70 ESCC	82 normal	Saliva, brush of thoracic esophagus	16S rRNA, NGS	Decreased as disease progressed	*Granulicatella, Rothia, Streptococcus, Gemella, Leptotrichia* and *Schaalia* common in patients with LGD. *Lactobacillus* common in patients with HGD. *Bosea, Solobacterium, Gemella*, and *Peptostreptococcus* common in patients with ESCC.
Jiang et al. (2021) [45] PMID: 34858881	Case–control	China, Jiangsu province	32 ESCC, 15 esophagitis	21 healthy controls	Esophageal biopsy	16S rRNA, NGS	Higher richness of microbiota in cases compared with controls	In ESCC, decreased *Fusobacteria* at phylum level and *Faecalibacterium, Bacteroides, Curvibacter*, and *Blautia* at genus level compared with controls.

Li, Zhen et al. (2021) [39] PMID: 34717543	Cross-sectional	China, Zhengzhou (Henan province)	41 ESCC tumor tissue (paired with non-tumor tissue), 70 individual ESCC	non-tumor tissue from 41 patients	Surgical specimens	16S rRNA, NGS, whole exome sequencing	Decreased in tumor tissue vs. non-tumor tissue	*Bacteroidetes, Fusobacteria,* and *Spirochaetae* were the top three significantly enriched taxa in the tumor group compared with normal group. *Butyrivibrio* and *Lactobacillus* lower in relative abundances in the tumor. *Streptococcus* and *Fusobacterium* increased in relative abundance in tumor groups.
Kovaleva et al. (2021) [43] PMID: 34203319	Cross-sectional	Russia	48 ESCC tumors	Matched controls normal epithelial cells	Surgical specimens	16S rRNA, NGS	No difference between tumor and normal tissue; depends on tumor stroma type (tumors with high CD8 + T cells have significant decrease in alpha diversity)	Relative abundance of *Staphylococcus* higher in tumor tissue than normal tissue.

Continued

Table 1 Summary of studies characterizing esophageal microbiome in ESCC—cont'd

Author (year)	Study Design	Country	Cases (N)	Controls (N)	Sample Type	Method	Alpha Diversity	Differentially Abundant Taxa
Guo et al. (2021) [46] PMID: 34955686	Cross-sectional	China, Henan province	370 ESCC patients	Paired nontumor tissue from 370 ESCC patients	Surgical specimens	Immunohistochemistry	Not assessed	*P. gingivalis* infection negatively correlated with Beclin1 expression in ESCC tissues. *P. gingivalis* infection and low Beclin1 expression associated with differentiation status, tumor invasion depth, lymph node metastasis, clinical stage and prognosis in ESCC patients.
Rao et al. (2021) [44] PMID: 33673801	Retrospective cohort	China, Fujian province	60 ESCC patients, drinkers	60 ESCC patients, nondrinkers	Surgical specimens	16S rRNA, NGS	Shannon diversity index higher in nondrinkers	Alcohol-related bacteria: *Clostridium*, *Helicobacter*, *Catonella*, *Bacteroides*, *Bacillus*, *Moraxella*, *Bulleidia*. Relative abundance of *Catonella* decreased with increasing frequency and years of alcohol consumption.

Li, D. et al. (2020) [37] PMID: 32676460	Case–control	China, Wuhan	17 ESCC, 11 EAC, 15 post-ESCC	16 healthy controls	Esophageal biopsy	16S rRNA, NGS	Lower diversity in ESCC than control, EAC than control, and post-ESCC than control. No difference between ESCC and EAC.	Relative abundance of *Fusobacteria* increased and *Actinobacteria* decreased in ESCC compared with control. Higher abundance of *Firmicutes* in EAC compared with ESCC.
Li, M. et al. (2020) [47] PMID: 33446998	Case–control	China, Linzhou, Henan province	70 esophagitis, 70 low-grade intraepithelial neoplasia (LGIN), 19 high-grade intraepithelial neoplasia (HGIN), and 7 ESCC	70 healthy controls	Esophageal brush, esophageal biopsy	16S rRNA, NGS	No difference between groups	In biopsy and swab specimens, decreasing *Firmicutes* and increasing *Fusobacteria* with the progression of ESCC. In biopsy samples, increasing *Neisseria* from normal to ESCC. Combination of *Streptococcus* and *Neisseria* best predicted ESCC and its precancerous lesions.

Continued

Table 1 Summary of studies characterizing esophageal microbiome in ESCC—cont'd

Author (year)	Study Design	Country	Cases (N)	Controls (N)	Sample Type	Method	Alpha Diversity	Differentially Abundant Taxa
Shao et al. (2019) [42] PMID: 31355925	Cross-sectional	China, Linzhou, Henan province	61 ESCC tumors, 34 gastric cardia adenocarcinoma (GCA) tumors	52 ESCC paired non-tumor tissue, 24 GCA paired non-tumor tissue	Surgical specimens	16S rRNA, NGS	No significant differences between ESCC and control, no association with tumor stage	Higher relative abundance of *Fusobacteria* and lower relative abundance of *Firmicutes* in ESCC compared with control. Lower relative abundance of *Streptococcus* in ESCC compared with control.
Liu et al. (2018) [41] PMID: 29923815	Retrospective cohort	China, Fujian province	45 ESCC	None (compared within ESCC group based on tumor stage and lymph node metastasis)	Surgical specimens	16S rRNA, NGS	No significant differences in alpha diversity based on tumor stage or lymph node involvement	Higher tumor stage group had increased abundance of *Streptococcus* compared with lower tumor stage group. Positive lymph node group had increased *Firmicutes*, *Bacteroidetes*, and *Spirochaetes*; and lower *Proteobacteria* compared with negative lymph node group.

								Positive lymph node group had higher abundance of *Prevotella* and *Treponema* compared with negative lymph node group. Abundances of genera *Prevotella* and *Streptococcus* were inversely correlated with survival rates in ESCC patients.
Yu et al. (2014) [38] PMID: 24700175	Cross-sectional	China, Linxian, Henan province	142 ESD	191 healthy controls	Esophageal sample by balloon	Human Oral Microbe Identification Microarray	Microbial richness lower in ESD	None

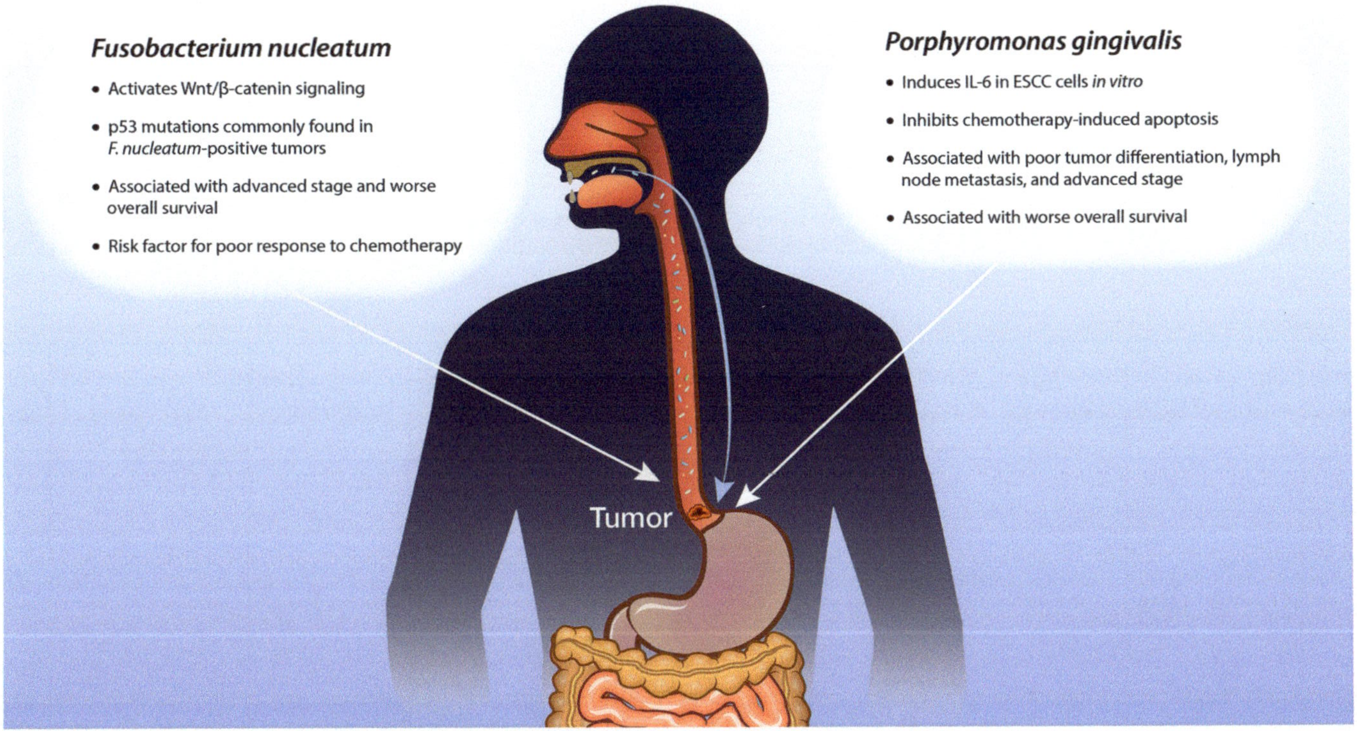

Fig. 1 Key oral pathogens associated with esophageal squamous cell carcinoma.

A study in the Henan province of China found that *P. gingivalis* was more commonly detected in ESCC compared with adjacent tissue or healthy tissue [51]. *P. gingivalis* presence was not associated with age, gender, or smoking history, but was positively associated with poor tumor differentiation, lymph node metastasis, and late TNM Classification of Malignant Tumors (TNM) stage, suggesting a role for the pathogen in tumor behavior. *P. gingivalis* presence was also associated with lower mean survival time (20 months for *P. gingivalis* positive vs. 26 months for *P. gingivalis* negative) [51]. The same group subsequently found that the degree of *P. gingivalis* infection is low in noncancerous tissues, increased in esophageal dysplasia, and at its highest in ESCC, suggesting that it could be related to esophageal cancer progression, or at least a marker of it [52].

A study in Taiwan corroborated these findings using human tissues and mouse models [53]. The investigators found that *P. gingivalis* was enriched in oral biofilms and esophageal tissue samples of ESCC patients and that infection was correlated with poor differentiation, advanced disease, higher locoregional recurrences, distant metastases, and reduced overall survival rates (adjusted HR 2.14, 95% CI 1.17–3.92 for *P. gingivalis* positive vs. negative). In mouse models of ESCC with *P. gingivalis* infection, *P. gingivalis* was associated with increased tumor growth and likelihood of developing invasive carcinoma via upregulation of the epithelial–mesenchymal transition. In vitro studies of human ESCC cell lines show *P. gingivalis* increases invasive potential by inducing IL-6 production, which induces expression of epithelial–mesenchymal transition markers and cancer stem cell markers. This theory is supported by elevated levels of IL-6 seen in primary ESCC specimens with *P. gingivalis* infection [53].

P. gingivalis may mediate not only progression of esophageal cancer, but also response to treatment. *P. gingivalis*positive ESCC patients had a lower response rate to chemotherapy and lower median 5-year overall survival (5-year survival rate 13.1% and median survival time 22 months in *P. gingivalis* positive vs. 30.1% and 35 months in *P. gingivalis* negative) [46]. *P. gingivalis* reduces chemotherapy-induced apoptosis and promotes proliferation of esophageal cancer cells via decreased Beclin1 expression and regulation of the STAT3 and caspase-3 proteins, all of which are involved in programmed cell death [46,54]. *P. gingivalis* may also be a possible target in therapy; metronidazole treatment after inoculation of mice with *P. gingivalis*-infected human esophageal cancer cells led to significantly decreased tumor volume [54].

F. nucleatum

Like *P. gingivalis*, *F. nucleatum* has been associated with periodontitis [55]. *F. nucleatum* is also known to be involved in colorectal carcinogenesis. *F. nucleatum* attaches to and invades colorectal cancer cells, subsequently activating the WNT/β-catenin signaling pathway and promoting tumor growth [56]. A clinical trial in Shanghai is currently exploring whether oral metronidazole can improve the efficacy of adjuvant chemotherapy in colorectal cancer [57].

A study in Japan found that *F. nucleatum* was present in 23% of ESCC cases and enriched in cancer tissues compared with adjacent nontumor tissues. *F. nucleatum* was not associated with age, sex, smoking, alcohol, tumor location, or tumor size, but was associated with more advanced tumor stage and higher cancer-specific mortality (adjusted HR 1.98, 95% CI 1.14–3.37) [58]. In addition, *F. nucleatum* may be a potential biomarker for response to chemotherapy. High *F. nucleatum* levels were an independent risk factor associated with poor recurrence-free survival, even at early tumor stages, and poor chemotherapeutic response [59]. A study in the Henan province of China found similar results, with *F. nucleatum* in 70% of ESCC tumor samples and at significantly higher levels compared with paired normal tissues. Assessment of a subset of *F. nucleatum*-positive samples for mutations showed common TP53 mutations along with mutations that may be unique to *F. nucleatum*-positive ESCC [39]. TP53 mutations have been detected in 59%–91% of ESCC patients and have been associated with poor response to preoperative chemotherapy [60–63]. *F. nucleatum* has been shown to affect the epigenome in human carotid artery endothelial cells [64]; further study is required to investigate the impact of *F. nucleatum* on host genome and epigenome in esophageal carcinogenesis.

Esophageal microbiome: Environmental exposures and carcinogen metabolism

The relationship between dietary intake and ESCC has been widely studied. In addition to possible direct carcinogenic or protective effects, dietary intake may play a role in ESCC carcinogenesis by altering the microbiome. Diet may modify risk of ESCC directly through changes in the esophageal microbiome and indirectly through changes in the distal gut microbiome by way of systemic effects and inflammation [65]. Fruit and vegetable intake has an inverse relationship with risk of ESCC [13,16,19,66,67], and red meat intake has a positive relationship with risk of ESCC [68–70], similar to other gastrointestinal cancers. Fruit and vegetable consumption has been associated with changes in microbial composition that result in increased production of antiinflammatory short-chain fatty acids in the colon [71,72]. Increased fiber intake, often in the form of fruits and vegetables, has also been associated with a distinct esophageal microbiome, though the downstream changes in esophageal microbial metabolism are not well understood [73].

The two most widely established carcinogens in ESCC are acetaldehyde and *N*-nitrosamines. Acetaldehyde is a byproduct of both host and microbe ethanol metabolism. Host metabolism occurs both locally, in the upper gastrointestinal tract, and systemically, in the liver. Esophageal mucosal alcohol dehydrogenase has 7–12 times the alcohol dehydrogenase activity of the tongue and gingiva and is the primary source of esophageal acetaldehyde. Esophageal mucosal acetaldehyde dehydrogenase activity further metabolizes acetaldehyde into acetic acid. Acetaldehyde dehydrogenase-2 (ALDH2) deficiency is present in approximately 50% of East Asians and dose-dependently increases the risk for esophageal cancer in alcohol drinkers [74,75].

Esophageal dysbiosis in ESCC may contribute to cancer development through altered microbial nitrate metabolism and increased production of *N*-nitrosamines [9]. In a multiprovincial study in China, predicted functional profiling analyses of oral and esophageal microbiome data in ESCC patients were notable for significant differences in nitrate and nitrite reductase signatures. In data from esophageal samples, the investigators found that nitrite reductase decreased significantly in ESCC and high-grade dysplasia compared with normal controls [40]. These analyses must be interpreted with caution, however, as they do not contain information on expression of genes in microbial samples. Alterations in nitrate metabolism and other metabolic pathways deserve further study.

Oral microbiome, Oral health, and Esophageal squamous cell carcinoma

Poor oral health has been established as a risk factor in ESCC in multiple regions of high incidence, including Iran, China, India, eastern Africa, Japan, central Europe, and Latin America [29,76–88]. To date, it has not been well studied in areas of lower incidence like the United States. Poor oral health is defined variably as inadequate oral hygiene habits, tooth loss, tooth decay, or periodontal disease. Analyses of the prospective Linxian Nutrition Intervention Trial Cohort have shown an association between tooth loss and increased risk of esophageal and gastric cancers, but not other cancers (including colorectal and liver cancer) [79,81,85]. Similarly, tooth brushing has been shown to be associated with a decreased risk of ESCC [29,80,82–84]. In eastern Africa, dental fluorosis has been recently identified as a risk factor for ESCC [29]. The mechanism of association between poor oral health and ESCC is not well understood, but it is hypothesized that the oral microbiome plays a role, as oral bacteria are known to be involved in periodontitis and dental caries [33,89].

Oral flora may contribute directly to ESCC through the initial local activation of carcinogens that pass through the upper aero-digestive tract. There is heavy overlap between the oral and esophageal microbiomes; key producers of acetaldehyde, including *Streptococcus*, *Neisseria*, *Rothia*, and *Prevotella*, are present in the esophageal microbiome and often enriched in ESCC [39–41,47,75]. These bacteria produce mutagenic levels of acetaldehyde in saliva in a dose-dependent fashion upon ingestion of ethanol [90–92]. Oral bacteria also play a role in activation of nitrosamines in tobacco smoke [93]. A near complete elimination of nitrate-reducing oral bacteria using antiseptic chlorhexidine mouthwash led to a 76% decrease in the amount of salivary nitrate that was produced in vivo [94].

Oral health may also contribute to ESCC carcinogenesis through increased systemic inflammation. Associations have been observed between oral health and pancreatic cancer [95,96]. Oral bacteria serve as the source of repeated transient systemic bacteremia after routine and iatrogenic traumas like tooth brushing, mastication, and dental procedures [97–99]. Periodontal disease, which has been associated with ESCC in addition to other cancers, further increases translocation of oral microbes into the bloodstream. These processes result in chronic systemic exposure to oral bacteria and an associated immune response and inflammation, which is felt to be carcinogenic [33,93].

Changes in the oral microbiome alter host esophageal mucosal gene expression and may impact carcinogenesis. A randomized controlled trial in the United States found that use of an antiseptic mouth rinse significantly altered the esophageal microbiome and was associated with marked reductions in esophageal tissue expression of periostin, claudin-18, histidine decarboxylase, chemokines CXCL1 and CXCL13, and several tumor necrosis factor receptors. Several of these genes are associated with inflammation, suggesting a potential link between the microbiome and esophageal disease [34].

Changes in the oral microbiome may also serve as a biomarker for ESCC, both at the diagnostic and therapeutic phases of care. Microbiome samples from saliva and oral swabs both correlate closely with samples from esophageal brushings [34,40]. Oral microbiome sampling methods are much less invasive than esophageal sampling, increasing their potential utility in screening and prognosticating. Several cross-sectional studies have assessed the oral microbiome in ESCC with inconsistent results. Alpha diversity in the oral microbiome of ESCC cases has been found to decrease [100], increase [101], or have no significant differences compared with controls [102–104]. Certain bacteria are enriched in the oral microbiome of ESCC patients, including genera *Prevotella*, *Streptococcus*, and *Porphyromonas* [100,104,105]. A study in China assessed salivary samples from patients with and without squamous dysplasia and ESCC and found microbiome alterations across the spectrum of progression, with notable increases in the genus *Streptococcus* [40]. Two prospective case control studies have identified oral bacteria associated with ESCC. A nested case–control study in China collected prediagnostic oral swabs and found 11 species that were predictive of severe esophageal squamous dysplasia or more advanced disease, including *F. nucleatum*, with adjusted odds ratios ranging from 1.98 to 10.93 [101]. A nested case–control study in the United States collected prediagnostic mouthwash samples and found that *P. gingivalis* was marginally associated with ESCC risk (odds ratio 1.30) [102]. Additional prospective studies testing predictive performance of possible microbiome biomarkers across different populations are needed.

Role of the gastric microenvironment in Esophageal squamous cell carcinoma

The gastric microbiome shares common dietary exposures with the esophageal microbiome and may directly influence the esophageal luminal environment in the setting of gastroesophageal reflux. Increased pH in the setting of atrophic gastritis is associated with nearly double the risk of ESCC [106], possibly mediated by alterations in gastric microbial composition that increase the production of acetaldehyde and *N*-nitrosamines [9]. Patients with atrophic gastritis have significantly elevated levels of intragastric acetaldehyde after ethanol exposure as well as increased gastric nitrosamines [107,108]. Although *Helicobacter pylori* is a major cause of atrophic gastritis, in a meta-analysis, it has been associated with decreased risk of ESCC in Eastern, but not Western populations [109]. Proton pump inhibitor (PPI) use, which also increases gastric pH, is associated with anaerobic bacterial overgrowth in the stomach and a significant increase in gastric acetaldehyde levels after ethanol ingestion [110]. Observational data from Sweden suggests that PPI use is associated with increased risk of ESCC [111,112]. However, studies on *H. pylori* and PPI use must be interpreted with caution as these exposures often associate with smoking, alcohol use, and other risk factors for ESCC [9].

The association of the above gastric conditions with increased risk of ESCC implies a potential mediating role of the gastric microbiome. Changes in the gastric microbiome have been noted in ESCC patients. A cross-sectional study in the Golestan province of Iran found that the composition of gastric corpus mucosal microbiota differs in ESCC and esophageal squamous dysplasia compared with healthy controls; *Clostridiales* and *Erysipelotrichales* taxa were increased in relative abundance in ESCC and esophageal squamous dysplasia [113]. Further study is necessary to elucidate the significance of these alterations.

Limitations of current research

The conclusions we are able to draw from current microbiome and ESCC studies are limited, primarily due to study design. Many studies are cross-sectional or do not adequately adjust for confounders. This combined with a lack of experimental data to validate associations seen at a population-level make it difficult to determine causality at this time. To what extent changes in the microbiome contribute to ESCC carcinogenesis or cause alterations in response to treatment remains unclear. Furthermore, in vivo studies and incorporation of microbiome sampling into cohort designs may further our understanding of causative roles of the esophageal microbiome.

Sampling variability poses another limitation on interpretation of existing research. Sampling techniques currently vary considerably in source and method, limiting the external generalizability of findings obtained with such techniques. Significant differences in microbial composition across geographic regions, age groups, and diets also restrict the generalizability of microbiome studies across populations [114]. Current research is heavily based in China, as well as the United States, Iran, India, and southeastern Africa. Future studies must standardize sample practices and obtain microbiome sample across all populations of interest.

Microbiome characterization techniques have advanced from microarray technology to 16S rRNA-based high-throughput DNA sequencing, the most common methodology in current esophageal microbiome studies. However, this method results in limited granularity on microbiome composition and is unable to provide direct information about bacterial species and strains as well as gene composition, which limits our understanding of the function of the microbiome in disease states. Shotgun metagenomic sequencing is a more expensive technique that has been used to sequence all DNA in samples; this allows profiling of nonbacterial members of the microbiome as well as evidence of gene functional variation. However, human DNA cannot be easily separated in this process. This poses an issue for metagenomic studies of the esophageal microbiome, as upper GI sampling techniques invariably contain large amounts of human DNA, on the order of 100,000:1 human:bacterial DNA in esophageal tissue biopsies and 100:1 in esophageal brush samples, which then yield only a small amount of microbial DNA for analysis [115]. Currently, metagenomics yields significantly different taxonomic results from 16S rRNA-based techniques, which may reflect higher sensitivity or higher error rate from lack of specificity [116,117]. The abundance of human DNA in upper GI samples confounds bacterial detection in mapping whole genome sequencing results to bacterial genomes [118]. Improvement in human and extracellular DNA depletion before metagenomic analysis or improvements in the computational pipeline for microbial detection from whole genome sequencing data may increase the value of metagenomic studies [118,119].

Future studies may also wish to expand in scope by assessing the contribution of both bacterial and nonbacterial microorganisms to carcinogenic metabolite production, such as acetaldehyde and *N*-nitrosamines. Oral and gastric microbiomes have been observed to contribute significantly to carcinogen exposure through acetaldehyde and nitrosamine production [9,90–93]. In addition to bacteria, *Candida* yeasts are a major contributor to acetaldehyde exposure, and chronic candidiasis has been linked to ESCC [75,120]. Recent studies have suggested that the esophageal microbiome in ESCC may have alterations in nitrate and nitrite metabolism [40].

Conclusions

Alterations in the oral, esophageal, and gastric microbiomes are associated with ESCC. These alterations may impact ESCC carcinogenesis through changes in the metabolism of acetaldehyde and *N*-nitrosamines and increased systemic inflammation. Certain bacteria such as *P. gingivalis* and *F. nucleatum* may be involved in ESCC tumor progression and could serve as prognostic markers. Noninvasive sampling of the oral microbiome may yield information about risk of ESCC as well. Overcoming the limitations of current research will require better consideration of confounding factors, increased prospective and experimental study design, and tailored use of metagenomic techniques.

Future directions

- Definition of the best microbial sampling and characterization techniques for oral, esophageal, and gastric microbial analysis.
- Weighted importance of diet and environmental factors on dysbiosis and specific risk for BE and ESCC.
- Metabolite production by the esophageal microbiome and mycobiome as a potential carcinogenic mechanisms.
- Mitigation strategies for risk reduction of oral, esophageal, and gastric dysbiosis related to BE and ESCC will need definition and prospective evaluation.

COI disclosures

Alyyah Malick, Ali Soroush, and Julian A. Abrams have no conflicts of interest to disclose.

References

[1] H. Sung, J. Ferlay, R.L. Siegel, et al., Global cancer statistics 2020: GLOBOCAN estimates of incidence and mortality worldwide for 36 cancers in 185 countries, CA Cancer J. Clin. 71 (3) (2021) 209–249.
[2] B. Njei, T.R. McCarty, J.W. Birk, Trends in esophageal cancer survival in United States adults from 1973 to 2009: a SEER database analysis, J. Gastroenterol. Hepatol. 31 (6) (2016) 1141–1146.
[3] SEER Cancer Statistics Review 1975–2011, 2014.
[4] H. He, N. Chen, Y. Hou, et al., Trends in the incidence and survival of patients with esophageal cancer: A SEER database analysis, Thorac Cancer. 11 (5) (2020) 1121–1128.
[5] M. Arnold, M. Laversanne, L.M. Brown, S.S. Devesa, F. Bray, Predicting the future burden of esophageal cancer by histological subtype: international trends in incidence up to 2030, Am. J. Gastroenterol. 112 (8) (2017) 1247–1255.
[6] Q.-L. Wang, S.-H. Xie, K. Wahlin, J. Lagergren, Global time trends in the incidence of esophageal squamous cell carcinoma, Clin. Epidemiol. 10 (2018) 717.
[7] H. Liang, J.-H. Fan, Y.-L. Qiao, Epidemiology, etiology, and prevention of esophageal squamous cell carcinoma in China, Cancer Biol. Med. 14 (1) (2017) 33.
[8] M. Hosseini-Bensenjan, H.M. Vardanjani, S. Haghpanah, Z. Khosravizadegan, K. Bagheri-Lankarani, Investigating Trends of Incidence Rates of Esophageal Cancer

Divided by Squamous Cell Carcinoma and Adenocarcinoma in Southern Iran: a 10-Year Experience, J. Gastrointest. Cancer 53 (1) (2022) 230–234.
[9] A. Soroush, A. Etemadi, J.A. Abrams, Non-acid fluid exposure and esophageal squamous cell carcinoma, Dig. Dis. Sci. 1–9 (2021).
[10] M. Matejcic, M.J. Gunter, P. Ferrari, Alcohol metabolism and oesophageal cancer: a systematic review of the evidence, Carcinogenesis 38 (9) (2017) 859–872.
[11] S.S. Hecht, Tobacco carcinogens, their biomarkers and tobacco-induced cancer, Nat. Rev. Cancer 3 (10) (2003) 733–744.
[12] A. Prabhu, K.O. Obi, J.H. Rubenstein, The synergistic effects of alcohol and tobacco consumption on the risk of esophageal squamous cell carcinoma: a meta-analysis, Am. J. Gastroenterol. 109 (6) (2014) 822–827.
[13] L.S. Engel, W.H. Chow, T.L. Vaughan, et al., Population attributable risks of esophageal and gastric cancers, J. Natl. Cancer Inst. 95 (18) (2003) 1404–1413.
[14] N. Pandeya, C.M. Olsen, D.C. Whiteman, Sex differences in the proportion of esophageal squamous cell carcinoma cases attributable to tobacco smoking and alcohol consumption, Cancer Epidemiol. 37 (5) (2013) 579–584.
[15] L. Morris Brown, R. Hoover, D. Silverman, et al., Excess incidence of squamous cell esophageal cancer among US Black men: role of social class and other risk factors, Am. J. Epidemiol. 153 (2) (2001) 114–122.
[16] G.D. Tran, X.D. Sun, C.C. Abnet, et al., Prospective study of risk factors for esophageal and gastric cancers in the Linxian general population trial cohort in China, Int. J. Cancer 113 (3) (2005) 456–463.
[17] C.C. Abnet, M. Arnold, W.-Q. Wei, Epidemiology of esophageal squamous cell carcinoma, Gastroenterology 154 (2) (2018) 360–373.
[18] Y. Chen, Y. Tong, C. Yang, et al., Consumption of hot beverages and foods and the risk of esophageal cancer: a meta-analysis of observational studies, BMC Cancer 15 (1) (2015) 1–13.
[19] M. Sheikh, H. Poustchi, A. Pourshams, et al., Individual and combined effects of environmental risk factors for esophageal cancer based on results from the Golestan Cohort Study, Gastroenterology 156 (5) (2019) 1416–1427.
[20] F. Islami, J. Ren, P. Taylor, F. Kamangar, Pickled vegetables and the risk of oesophageal cancer: a meta-analysis, Br. J. Cancer 101 (9) (2009) 1641–1647.
[21] N.D. Freedman, Y. Park, A.F. Subar, et al., Fruit and vegetable intake and esophageal cancer in a large prospective cohort study, Int. J. Cancer 121 (12) (2007) 2753–2760.
[22] S. Jeurnink, F. Büchner, H. Bueno-de-Mesquita, et al., Variety in vegetable and fruit consumption and the risk of gastric and esophageal cancer in the European prospective investigation into cancer and nutrition, Int. J. Cancer 131 (6) (2012) E963–E973.
[23] X. Cai, C. Wang, W. Yu, et al., Selenium exposure and cancer risk: an updated meta-analysis and meta-regression, Sci. Rep. 6 (1) (2016) 1–18.
[24] D. Nasrollahzadeh, F. Kamangar, K. Aghcheli, et al., Opium, tobacco, and alcohol use in relation to oesophageal squamous cell carcinoma in a high-risk area of Iran, Br. J. Cancer 98 (11) (2008) 1857–1863.
[25] M. Sheikh, R. Shakeri, H. Poustchi, et al., Opium use and subsequent incidence of cancer: results from the Golestan Cohort Study, Lancet Glob. Health 8 (5) (2020) e649–e660.
[26] C.P. Wen, M.K. Tsai, W.S.I. Chung, et al., Cancer risks from betel quid chewing beyond oral cancer: a multiple-site carcinogen when acting with smoking, Cancer Causes Control 21 (9) (2010) 1427–1435.
[27] S. Akhtar, A.A. Sheikh, H.U. Qureshi, Chewing areca nut, betel quid, oral snuff, cigarette smoking and the risk of oesophageal squamous-cell carcinoma in South Asians: a multicentre case–control study, Eur. J. Cancer 48 (5) (2012) 655–661.

[28] R. Gupta, L.C. Mariano, S.T. Nethan, et al., Risk reversal of oral, pharyngeal and oesophageal cancers after cessation of betel quid users: a systematic review and meta-analysis, Ann. Glob. Health 88 (1) (2022).
[29] D. Menya, S.K. Maina, C. Kibosia, et al., Dental fluorosis and oral health in the African Esophageal Cancer Corridor: Findings from the Kenya ESCCAPE case–control study and a pan-African perspective, Int. J. Cancer 145 (1) (2019) 99–109.
[30] S. Okello, S.J. Akello, E. Dwomoh, et al., Biomass fuel as a risk factor for esophageal squamous cell carcinoma: a systematic review and meta-analysis, Environ. Health 18 (1) (2019) 1–11.
[31] M.D. Gammon, H. Ahsan, J.B. Schoenberg, et al., Tobacco, alcohol, and socioeconomic status and adenocarcinomas of the esophagus and gastric cardia, J. Natl. Cancer Inst. 89 (17) (1997) 1277–1284.
[32] N.A. Dar, I.A. Shah, G.A. Bhat, et al., Socioeconomic status and esophageal squamous cell carcinoma risk in Kashmir, India, Cancer Sci. 104 (9) (2013) 1231–1236.
[33] Y. Yano, A. Etemadi, C.C. Abnet, Microbiome and cancers of the esophagus: a review, Microorganisms 9 (8) (2021) 1764.
[34] M.K. Annavajhala, M. May, G. Compres, et al., Relationship of the esophageal microbiome and tissue gene expression and links to the oral microbiome: a randomized clinical trial, Clin. Transl. Gastroenterol. 11 (12) (2020).
[35] R.F. Schwabe, C. Jobin, The microbiome and cancer, Nat. Rev. Cancer 13 (11) (2013) 800–812.
[36] W. Yang, C.-H. Chen, M. Jia, et al., Tumor-associated microbiota in esophageal squamous cell carcinoma, Front. Cell Dev. Biol. 9 (2021).
[37] D. Li, R. He, G. Hou, et al., Characterization of the esophageal microbiota and prediction of the metabolic pathways involved in esophageal cancer, Front. Cell. Infect. Microbiol. 268 (2020).
[38] G. Yu, M.H. Gail, J. Shi, et al., Association between upper digestive tract microbiota and cancer-predisposing states in the esophagus and stomach, Cancer Epidemiol. Biomarkers Prev. 23 (5) (2014) 735–741.
[39] Z. Li, C. Shi, J. Zheng, et al., *Fusobacterium nucleatum* predicts a high risk of metastasis for esophageal squamous cell carcinoma, BMC Microbiol. 21 (1) (2021) 1–14.
[40] Z. Li, L. Dou, Y. Zhang, et al., Characterization of the oral and esophageal microbiota in esophageal precancerous lesions and squamous cell carcinoma, Front. Cell. Infect. Microbiol. 11 (2021).
[41] Y. Liu, Z. Lin, Y. Lin, et al., Streptococcus and Prevotella are associated with the prognosis of oesophageal squamous cell carcinoma, J. Med. Microbiol. 67 (8) (2018) 1058–1068.
[42] D. Shao, E. Vogtmann, A. Liu, et al., Microbial characterization of esophageal squamous cell carcinoma and gastric cardia adenocarcinoma from a high-risk region of China, Cancer 125 (22) (2019) 3993–4002.
[43] O. Kovaleva, P. Podlesnaya, M. Rashidova, et al., Prognostic significance of the microbiome and stromal cells phenotype in esophagus squamous cell carcinoma, Biomedicine 9 (7) (2021) 743.
[44] W. Rao, Z. Lin, S. Liu, et al., Association between alcohol consumption and oesophageal microbiota in oesophageal squamous cell carcinoma, BMC Microbiol. 21 (1) (2021) 1–12.
[45] Z. Jiang, J. Wang, Z. Shen, Z. Zhang, S. Wang, Characterization of Esophageal Microbiota in Patients With Esophagitis and Esophageal Squamous Cell Carcinoma, Front Cell Infect Microbiol 11 (2021), 774330.
[46] Y. Guo, Y. Liu, H. Yang, et al., Associations of *Porphyromonas gingivalis* infection and low Beclin1 expression with clinicopathological parameters and survival of esophageal squamous cell carcinoma patients, Pathol. Oncol. Res. 27 (2021).

[47] M. Li, D. Shao, J. Zhou, et al., Signatures within esophageal microbiota with progression of esophageal squamous cell carcinoma, Chin. J. Cancer Res. 32 (6) (2020) 755.
[48] D. Michaud, K. Kelsey, E. Papathanasiou, C. Genco, E. Giovannucci, Periodontal disease and risk of all cancers among male never smokers: an updated analysis of the Health Professionals Follow-up Study, Ann. Oncol. 27 (5) (2016) 941–947.
[49] N.N. Nwizu, J.R. Marshall, K. Moysich, et al., Periodontal disease and incident cancer risk among postmenopausal women: results from the women's health initiative observational cohort, Cancer Epidemiol. Biomarkers Prev. 26 (8) (2017) 1255–1265.
[50] M. Costalonga, M.C. Herzberg, The oral microbiome and the immunobiology of periodontal disease and caries, Immunol. Lett. 162 (2) (2014) 22–38.
[51] S. Gao, S. Li, Z. Ma, et al., Presence of *Porphyromonas gingivalis* in esophagus and its association with the clinicopathological characteristics and survival in patients with esophageal cancer, Infect Agent Cancer 11 (1) (2016) 1–9.
[52] X. Yuan, Y. Liu, J. Kong, et al., Different frequencies of *Porphyromonas gingivalis* infection in cancers of the upper digestive tract, Cancer Lett. 404 (2017) 1–7.
[53] M.-F. Chen, M.-S. Lu, C.-C. Hsieh, W.-C. Chen, Porphyromonas gingivalis promotes tumor progression in esophageal squamous cell carcinoma, Cell. Oncol. 44 (2) (2021) 373–384.
[54] S. Gao, Y. Liu, X. Duan, et al., *Porphyromonas gingivalis* infection exacerbates oesophageal cancer and promotes resistance to neoadjuvant chemotherapy, Br. J. Cancer 125 (3) (2021) 433–444.
[55] A.L. Griffen, C.J. Beall, J.H. Campbell, et al., Distinct and complex bacterial profiles in human periodontitis and health revealed by 16S pyrosequencing, ISME J. 6 (6) (2012) 1176–1185.
[56] M.R. Rubinstein, X. Wang, W. Liu, Y. Hao, G. Cai, Y.W. Han, Fusobacterium nucleatum promotes colorectal carcinogenesis by modulating E-cadherin/β-catenin signaling via its FadA adhesin, Cell Host Microbe 14 (2) (2013) 195–206.
[57] Clinical Trials, n.d., Study of Oral Metronidazole on Postoperative Chemotherapy in Colorectal Cancer. In: https://ClinicalTrials.gov/show/NCT04264676.
[58] K. Yamamura, Y. Baba, S. Nakagawa, et al., Human microbiome *Fusobacterium nucleatum* in esophageal cancer tissue is associated with prognosis, Clin. Cancer Res. 22 (22) (2016) 5574–5581.
[59] K. Yamamura, D. Izumi, R. Kandimalla, et al., Intratumoral *Fusobacterium nucleatum* levels predict therapeutic response to neoadjuvant chemotherapy in esophageal squamous cell carcinoma, Clin. Cancer Res. 25 (20) (2019) 6170–6179.
[60] M. Izutsu, T. Domoto, S. Kamoshida, et al., Expression status of p53 and organic cation transporter 1 is correlated with poor response to preoperative chemotherapy in esophageal squamous cell carcinoma, World J. Surg. Oncol. 20 (1) (2022) 1–11.
[61] Y.-B. Gao, Z.-L. Chen, J.-G. Li, et al., Genetic landscape of esophageal squamous cell carcinoma, Nat. Genet. 46 (10) (2014) 1097–1102.
[62] C.G.A.R. Network, Integrated genomic characterization of oesophageal carcinoma, Nature 541 (7636) (2017) 169.
[63] S. Moody, S. Senkin, S. Islam, et al., Mutational signatures in esophageal squamous cell carcinoma from eight countries with varying incidence, Nat. Genet. 53 (11) (2021) 1553–1563.
[64] C.A. Despins, S.D. Brown, A.V. Robinson, A.J. Mungall, E. Allen-Vercoe, R.A. Holt, Modulation of the host cell transcriptome and epigenome by *Fusobacterium nucleatum*, MBio 12 (5) (2021), e0206221.
[65] Z. Al Bander, M.D. Nitert, A. Mousa, N. Naderpoor, The gut microbiota and inflammation: an overview, Int. J. Environ. Res. Public Health 17 (20) (2020) 7618.
[66] Y. Fan, J.-M. Yuan, R. Wang, Y.-T. Gao, M.C. Yu, Alcohol, tobacco, and diet in relation to esophageal cancer: the Shanghai Cohort Study, Nutr. Cancer 60 (3) (2008) 354–363.

[67] T. Yamaji, M. Inoue, S. Sasazuki, et al., Fruit and vegetable consumption and squamous cell carcinoma of the esophagus in Japan: the JPHC study, Int. J. Cancer 123 (8) (2008) 1935–1940.
[68] A. Golozar, A. Etemadi, F. Kamangar, et al., Food preparation methods, drinking water source and esophageal squamous cell carcinoma in the high risk area of golestan, Northeast Iran, Eur. J. Cancer Prev. 25 (2) (2016) 123.
[69] A. Keszei, L. Schouten, R. Goldbohm, P. Van Den Brandt, Red and processed meat consumption and the risk of esophageal and gastric cancer subtypes in The Netherlands Cohort Study, Ann. Oncol. 23 (9) (2012) 2319–2326.
[70] A.J. Cross, N.D. Freedman, J. Ren, et al., Meat consumption and risk of esophageal and gastric cancer in a large prospective study, Am. J. Gastroenterol. 106 (3) (2011) 432.
[71] C.H. Kim, J. Park, M. Kim, Gut microbiota-derived short-chain fatty acids, T cells, and inflammation, Immune Netw. 14 (6) (2014) 277–288.
[72] C. De Filippo, D. Cavalieri, M. Di Paola, et al., Impact of diet in shaping gut microbiota revealed by a comparative study in children from Europe and rural Africa, Proc. Natl. Acad. Sci. U. S. A. 107 (33) (2010) 14691–14696.
[73] Y.R. Nobel, E.J. Snider, G. Compres, et al., Increasing dietary fiber intake is associated with a distinct esophageal microbiome, Clin. Transl. Gastroenterol. 9 (10) (2018).
[74] S.-J. Yang, A. Yokoyama, T. Yokoyama, et al., Relationship between genetic polymorphisms of ALDH2 and ADH1B and esophageal cancer risk: a meta-analysis, World J. Gastroenterol. 16 (33) (2010) 4210.
[75] M.T. Nieminen, M. Salaspuro, Local acetaldehyde—an essential role in alcohol-related upper gastrointestinal tract carcinogenesis, Cancers (Basel) 10 (1) (2018) 11.
[76] N. Guha, P. Boffetta, V. Wünsch Filho, et al., Oral health and risk of squamous cell carcinoma of the head and neck and esophagus: results of two multicentric case-control studies, Am. J. Epidemiol. 166 (10) (2007) 1159–1173.
[77] A. Hiraki, K. Matsuo, T. Suzuki, T. Kawase, K. Tajima, Teeth loss and risk of cancer at 14 common sites in Japanese, Cancer Epidemiol. Biomarkers Prev. 17 (5) (2008) 1222–1227.
[78] K. Patel, J. Wakhisi, S. Mining, A. Mwangi, R. Patel, Esophageal cancer, the topmost cancer at MTRH in the Rift Valley, Kenya, and its potential risk factors, ISRN Oncol. 2013 (2013).
[79] Y. Yano, J. Fan, S.M. Dawsey, Y. Qiao, C.C. Abnet, A long-term follow-up analysis of associations between tooth loss and multiple cancers in the Linxian General Population cohort, J. Natl. Cancer Cent. 1 (2) (2021) 39–43.
[80] Y. Yano, C.C. Abnet, H. Poustchi, et al., Oral health and risk of upper gastrointestinal cancers in a large prospective study from a high-risk region: Golestan cohort study, Cancer Prev. Res. (Phila.) 14 (7) (2021) 709–718.
[81] C.C. Abnet, Y.-L. Qiao, S.D. Mark, Z.-W. Dong, P.R. Taylor, S.M. Dawsey, Prospective study of tooth loss and incident esophageal and gastric cancers in China, Cancer Causes Control 12 (9) (2001) 847–854.
[82] C.C. Abnet, F. Kamangar, F. Islami, et al., Tooth loss and lack of regular oral hygiene are associated with higher risk of esophageal squamous cell carcinoma, Cancer Epidemiol. Biomarkers Prev. 17 (11) (2008) 3062–3068.
[83] N.A. Dar, F. Islami, G. Bhat, et al., Poor oral hygiene and risk of esophageal squamous cell carcinoma in Kashmir, Br. J. Cancer 109 (5) (2013) 1367–1372.
[84] X. Chen, Z. Yuan, M. Lu, Y. Zhang, L. Jin, W. Ye, Poor oral health is associated with an increased risk of esophageal squamous cell carcinoma-a population-based case-control study in China, Int. J. Cancer 140 (3) (2017) 626–635.
[85] S. Zhang, P. Yu, J.B. Wang, J.H. Fan, Y.L. Qiao, P.R. Taylor, Association between tooth loss and upper gastrointestinal cancer: A 30-year follow-up of the Linxian Dysplasia Nutrition Intervention Trial Cohort, Thorac Cancer. 10 (4) (2019) 966–974.

[86] B.T. Mmbaga, A. Mwasamwaja, G. Mushi, et al., Missing and decayed teeth, oral hygiene and dental staining in relation to esophageal cancer risk: ESCCAPE case-control study in Kilimanjaro, Tanzania, Int. J. Cancer 148 (10) (2021) 2416–2428.
[87] A. Sepehr, F. Kamangar, S. Fahimi, F. Saidi, C.C. Abnet, S.M. Dawsey, Poor oral health as a risk factor for esophageal squamous dysplasia in northeastern Iran, Anticancer Res 25 (1B) (2005) 543–546.
[88] Y. Sun, T. Zhang, W. Wu, et al., Risk factors associated with precancerous lesions of esophageal squamous cell carcinoma: A screening study in a high risk Chinese population, J. Cancer 10 (14) (2019) 3284.
[89] R.J. Lamont, H. Koo, G. Hajishengallis, The oral microbiota: dynamic communities and host interactions, Nat. Rev. Microbiol. 16 (12) (2018) 745–759.
[90] N. Homann, H. Jousimies-Somer, K. Jokelainen, R. Heine, M. Salaspuro, High acetaldehyde levels in saliva after ethanol consumption: methodological aspects and pathogenetic implications, Carcinogenesis 18 (9) (1997) 1739–1743.
[91] M. Muto, Y. Hitomi, A. Ohtsu, et al., Acetaldehyde production by non-pathogenic Neisseria in human oral microflora: implications for carcinogenesis in upper aerodigestive tract, Int. J. Cancer 88 (3) (2000) 342–350.
[92] J. Kurkivuori, V. Salaspuro, P. Kaihovaara, et al., Acetaldehyde production from ethanol by oral streptococci, Oral Oncol. 43 (2) (2007) 181–186.
[93] J. Ahn, C.Y. Chen, R.B. Hayes, Oral microbiome and oral and gastrointestinal cancer risk, Cancer Causes Control 23 (3) (2012) 399–404.
[94] K. Shapiro, J. Hotchkiss, D. Roe, Quantitative relationship between oral nitrate-reducing activity and the endogenous formation of N-nitrosoamino acids in humans, Food Chem. Toxicol. 29 (11) (1991) 751–755.
[95] D.S. Michaud, K. Joshipura, E. Giovannucci, C.S. Fuchs, A prospective study of periodontal disease and pancreatic cancer in US male health professionals, J. Natl. Cancer Inst. 99 (2) (2007) 171–175.
[96] R.Z. Stolzenberg-Solomon, K.W. Dodd, M.J. Blaser, J. Virtamo, P.R. Taylor, D. Albanes, Tooth loss, pancreatic cancer, and *Helicobacter pylori*, Am. J. Clin. Nutr. 78 (1) (2003) 176–181.
[97] K. Crasta, C.G. Daly, D. Mitchell, B. Curtis, D. Stewart, L.J. Heitz-Mayfield, Bacteraemia due to dental flossing, J. Clin. Periodontol. 36 (4) (2009) 323–332.
[98] F.K. Bahrani-Mougeot, B.J. Paster, S. Coleman, J. Ashar, S. Barbuto, P.B. Lockhart, Diverse and novel oral bacterial species in blood following dental procedures, J. Clin. Microbiol. 46 (6) (2008) 2129–2132.
[99] P.B. Lockhart, M.T. Brennan, H.C. Sasser, P.C. Fox, B.J. Paster, F.K. Bahrani-Mougeot, Bacteremia associated with toothbrushing and dental extraction, Circulation 117 (24) (2008) 3118–3125.
[100] X. Chen, B. Winckler, M. Lu, et al., Oral microbiota and risk for esophageal squamous cell carcinoma in a high-risk area of China, PLoS One 10 (12) (2015), e0143603.
[101] F. Liu, M. Liu, Y. Liu, et al., Oral microbiome and risk of malignant esophageal lesions in a high-risk area of China: A nested case-control study, Chin. J. Cancer Res. 32 (6) (2020) 742.
[102] B.A. Peters, J. Wu, Z. Pei, et al., Oral microbiome composition reflects prospective risk for esophageal cancers, Cancer Res. 77 (23) (2017) 6777–6787.
[103] Q. Wang, Y. Rao, X. Guo, et al., Oral microbiome in patients with oesophageal squamous cell carcinoma, Sci. Rep. 9 (1) (2019) 1–9.
[104] Q. Zhao, T. Yang, Y. Yan, et al., Alterations of oral microbiota in Chinese patients with esophageal cancer, Front. Cell. Infect. Microbiol. 563 (2020).
[105] H. Li, Z. Luo, H. Zhang, et al., Characteristics of oral microbiota in patients with esophageal cancer in China, Biomed. Res. Int. 2021 (2021).

[106] F. Islami, P. Sheikhattari, J. Ren, F. Kamangar, Gastric atrophy and risk of oesophageal cancer and gastric cardia adenocarcinoma—a systematic review and meta-analysis, Ann. Oncol. 22 (4) (2011) 754–760.
[107] J. Kobayashi, Effect of diet and gut environment on the gastrointestinal formation of N-nitroso compounds: a review, Nitric Oxide 73 (2018) 66–73.
[108] S. Väkeväinen, S. Mentula, H. Nuutinen, et al., Ethanol-derived microbial production of carcinogenic acetaldehyde in achlorhydric atrophic gastritis, Scand. J. Gastroenterol. 37 (6) (2002) 648–655.
[109] F.-J. Xie, Y.-P. Zhang, Q.-Q. Zheng, et al., *Helicobacter pylori* infection and esophageal cancer risk: an updated meta-analysis, World J. Gastroenterol. 19 (36) (2013) 6098.
[110] S. Väkeväinen, J. Tillonen, M. Salaspuro, H. Jousimies-Somer, H. Nuutinen, M. Färkkilä, Hypochlorhydria induced by a proton pump inhibitor leads to intragastric microbial production of acetaldehyde from ethanol, Aliment. Pharmacol. Ther. 14 (11) (2000) 1511–1518.
[111] N. Brusselaers, L. Engstrand, J. Lagergren, Maintenance proton pump inhibition therapy and risk of oesophageal cancer, Cancer Epidemiol. 53 (2018) 172–177.
[112] N. Brusselaers, J. Lagergren, L. Engstrand, Duration of use of proton pump inhibitors and the risk of gastric and oesophageal cancer, Cancer Epidemiol. 62 (2019), 101585.
[113] D. Nasrollahzadeh, R. Malekzadeh, A. Ploner, et al., Variations of gastric corpus microbiota are associated with early esophageal squamous cell carcinoma and squamous dysplasia, Sci. Rep. 5 (1) (2015) 1–6.
[114] C.H. Park, S.K. Lee, Exploring esophageal microbiomes in esophageal diseases: a systematic review, J. Neurogastroenterol. Motil. 26 (2) (2020) 171.
[115] A. Gall, J. Fero, C. McCoy, et al., Bacterial composition of the human upper gastrointestinal tract microbiome is dynamic and associated with genomic instability in a Barrett's esophagus cohort, PLoS One 10 (6) (2015), e0129055.
[116] D. Peterson, K.S. Bonham, S. Rowland, C.W. Pattanayak, V. Klepac-Ceraj, Comparative analysis of 16S rRNA gene and metagenome sequencing in pediatric gut microbiomes, Front. Microbiol. 12 (2021) 1651.
[117] F. Durazzi, C. Sala, G. Castellani, G. Manfreda, D. Remondini, A. De Cesare, Comparison between 16S rRNA and shotgun sequencing data for the taxonomic characterization of the gut microbiota, Sci. Rep. 11 (1) (2021) 1–10.
[118] C. Zhang, K. Cleveland, F. Schnoll-Sussman, et al., Identification of low abundance microbiome in clinical samples using whole genome sequencing, Genome Biol. 16 (1) (2015) 1–16.
[119] M.T. Nelson, C.E. Pope, R.L. Marsh, et al., Human and extracellular DNA depletion for metagenomic analysis of complex clinical infection samples yields optimized viable microbiome profiles, Cell Rep. 26 (8) (2019). 2227–2240. e2225.
[120] C. Delsing, C. Bleeker-Rovers, F. van de Veerdonk, et al., Association of esophageal candidiasis and squamous cell carcinoma, Med. Mycol. Case Rep. 1 (1) (2012) 5–8.

CHAPTER 8

Esophageal dysbiosis and eosinophilic esophagitis—Bacteria to blame?

David A. Katzka
Columbia University Medical School, Division of Digestive and Liver Disease, New York, NY, United States

Introduction

Virtually every surface of the human body has a microbiome. This microbiome is defined by the specific organisms that populate the surface (microbiota) and their interaction with characteristics of the surface they populate. The columnar lined mucus layer of the small intestine is enriched with specific bacteria that line the surface and lumen to contribute to key body functions such as immunity, metabolism, and defense needed in the gut. In contrast, the relatively dry squamous mucosa of the skin interacting with unique microbiota will lead to distinct functions native to this area such as protection from pathogens. In essence, this interaction of bacteria, fungi, and viruses with the body surface functionally comprise an organ in which the microbiota helps to maintain a functional homeostasis. Notably, this is localized to the surface but the effects of the microbial derivatives can stimulate proinflammatory cytokines that yield a panesophageal effect particularly in barrier function [1] and promotion of generalized inflammation [2]. In addition, perturbation of the surface epithelium, such as may occur in diseases that increase the permeability of the epithelial surface or enable colonization with abnormal and/or increased populations of specific microbial colonies can lead to disease states. Notably, these disturbances in microbiota can be formidable and diverse as the human microbiome is estimated to consist of trillions of organisms and up to 1000 different bacterial species. These abnormal gut populations are termed dysbiosis or an altered microbiome.

Normal esophageal microbiome

Characterizing the normal microbiome of the esophagus, as in any organ, is difficult. This difficulty arises from the dynamic nature of the microbial

Esophageal Disease and the Role of the Microbiome
https://doi.org/10.1016/B978-0-323-95070-1.00010-2

populations potentially sensitive to changes in diet, medication use, and host genetics. Nevertheless, several studies have characterized the normal findings. In one study from almost two decades ago, [3] the microbiota was presented by 41 genera, including most commonly gram-positive bacteria *Streptococcus, Prevotella,* and *Veillonella* (spp). Overall, 140 bacteria have been identified. The most common phyla were *Firmicutes, Bacteroides, Actinobacteria, Proteobacteria, Fusobacteria* (spp), and *T-M7.* In a study approximately one decade ago [4] of the oral cavity and the upper and lower esophagus, *Streptococcus viridans* was found in 98% of subjects followed by fusobacterium, Neisseria, Haemophilus, and Prevotella. Notably, findings in the proximal and distal esophagus were similar.

Potential modulators of the Esophageal microbiome in eosinophilic esophagitis

Several factors must be considered when studying dysbiosis in EoE. The first and perhaps most important, is that the esophageal epithelium in EoE is more permeable than normal. This increased permeability has been well established through several techniques. These include in vivo measurement of esophageal impedance [5] and histologic confirmation of dilation of intercellular spaces [6], abrogation tight junctions [6] and decreased levels of specific tight junction [6–8] and desmosomal proteins [9]. The underlying cause of this epithelial dysfunction may be genomic as identified through germ line differences in genes that code for specific aspects of epithelial integrity or epigenetic related to inflammation. For example, genome-wide association studies in patients with EoE support germ line abnormalities in functional genes such as filaggrin, [10] a key tight junction protein in esophageal epithelium. Similarly, desmoglein-1, an adhesion molecule that also joins cells, is decreased in patients with EoE [9]. The significance of this leaky esophageal epithelium is the predisposition to increased epithelial exposure to penetrating bacterial products and/or components from the microbiota and other antigens. As a result, EoE as a disease is related to both abnormal populations of surface bacteria and the ability of bacterial products or components to activate pathogenic pathways with epithelial entry. For example, one study demonstrated the presence of *Haemophilus influenza* lipopolysaccharide (an endotoxin) in the esophagus of active EoE (Ravi et al., preliminary results). Gut wall exposure to this highly antigenic bacterial product is implicated in the pathogenesis of other immunologic diseases such as Crohn's disease [11]. Furthermore, lipopolysaccharide (LPS) binds to thymic stromal lymphopoietin (TSLP) [12], a Th2 cytokine activated in EoE. The activation of TSLP appears to be a major contributor to inflammation by induction of Th2 differentiation [13]. One study has demonstrated an increase in *Haemophilus* (spp) in EoE patients with active EoE [14].

A most recent study demonstrated that microbial colonization has a direct effect on esophageal tissue inflammation and the epithelial integrity, as well expression of genes involved in EoE (including *POSTN, KLK5,* and *HIF1A*) [1]. A recent review on the esophageal microbiome effects on esophageal mucosal diseases has highlighted the potential pivotal role in the pathogenesis of EoE, interacting with genetic and environmental exposures [2].

One must also consider variation in diet (including fasting) and use of medications in EoE as potential modulators of the esophageal microbiome. In the gut, reduction of dietary constituents such as fiber is associated with a change in bacteria representation [15]. Use of proton pump inhibitors (PPIs) has also been demonstrated to alter the gut microbiota [16,17] as a potential predisposition to *Clostridium difficile* infection. As diet elimination and proton pump inhibitors are two of the primary therapies in eosinophilic esophagitis (EoE), changes in bacterial populations on the esophageal surface might be anticipated. Also, unlike the gut, the esophagus lacks a mucus layer as another means of attenuating entry of the microbiota on epithelial function. Whether this is effectively countered by the tighter syncytium of squamous epithelium when compared with columnar mucosa is unclear. A glycocalyx layer is also present providing further protection to the esophageal lining. Finally, a change in mouth/salivary flora might lead to a change in the esophageal microbiome. For example, *Firmicutes* and *Fusobacteria* (spp) are commonly found in both the mouth and esophagus, suggesting swallowed bacteria are an important source of esophageal microbiota [18,19]. Although the esophagus also has unique bacterial populations when compared with the mouth, this commonality suggests normal and pathologic influences on mouth bacteria may further affect esophageal function in patients with EoE [20].

Sampling methods used to study the microbiome in EoE

There has been variation in the esophageal collection techniques used to collect esophageal specimens for microbial analysis in EoE. These include biopsy and/or brushings [21,22] and scraping secretions from a string device [14]. This may be important as the string device collects bacteria localized to esophageal secretions, whereas brushing and biopsies may detect bacterial populations more intimately related to the esophageal epithelium. No direct comparisons have been performed.

EoE microbiome

Six studies have been performed to date studying the microbiome in patients with EoE. In a study by Harris et al., using the string device, an overall increase decrease in bacterial load was found in EoE when compared with normal subjects [14]. There was also a higher amount of *Haemophilus* (spp) in all untreated EoE subjects when compared with treated patients and controls. In principal component analysis, the EoE subjects were generally characterized by larger amounts *of Haemophilus, Pasteurella, Fusobacterium*, and *Aggregatibacter* (spp) and smaller amounts of *Actinomyces, Veillonella*, and *Rothia* (spp). Interpretation of this study is hampered by the previous potential for wide variability in studies; in this study, the bacterial population findings in untreated EoE patients were significantly different between the two collection centers in density of *Actinobacteria* and *Firmicutes (spp)*.

In a study by Benitez et al., using esophageal biopsies in pediatric EoE patients [22], proportions of bacterial communities were significantly different comparing all EoE esophageal microbiota to non-EoE controls, with enrichment of Proteobacteria, including Neisseria and Corynebacterium in the EoE cohort, and predominance of the Firmicutes in non-EoE control subjects. The *Streptococcus* and *Atopobium* genera were consistently enriched in non-EoE control samples. The use of PPIs in EoE patients did not affect the results. When samples were compared between actively inflamed EoE biopsies and non-EoE controls, there was a distinct shift in relative abundance of the esophageal microbiota, but not in community membership. Overall, targeted dietary intervention using a six-food elimination diet did not lead to significant differences in esophageal microbiota, whereas reintroduction of highly allergenic foods led to enrichment in *Granulicatella* and *Campylobacter* genera in the esophagus. In this study, oral and esophageal samples were also compared and found to have overlap of findings.

In a study by Norder Grussel et al., using brushings and biopsies [21], subjects with EoE had a significantly higher median number of bacterial species or groups similar to findings by Harris et al. This provides some reassurance in comparing the string collection to biopsy/brush. Alfa-streptococci (*Viridans streptococcus*, including species such as *Streptococcus salivarius, Streptococcus mutans, Streptococcus mitis, Streptococcus sanguinis*, and

Streptococcus anginosus) were predominant in both the proximal and distal esophagus. Other common bacteria were species of *Prevotella, Neisseria, Hemophilus, Fusobacterium* and *Lactobacillus* (spp). In comparison with healthy volunteers, patients with EoE had significantly more esophageal species and groups present within the esophagus. Streptococci were present in the EoE subjects throughout the esophagus.

In a comprehensive study from Hiremath et al., the EoE microbiome was assessed through collection of saliva. The data revealed that children with active EoE had a significantly lower relative abundance of Leptotrichiaceae unclassified and tended to have lower abundances of *Actinomyces*, *Lactobacillus*, and *Streptococcus* compared with non-EoE controls. The non-EoE controls had significantly higher relative abundance of Neisseriaceae compared with children with active EoE. A significantly higher relative abundance of *Haemophilus* was observed in children with active EoE when compared with children with inactive EoE. The relative abundance of *Haemophilus* and *Lactobacillus* significantly increased with increasing density of eosinophilic infiltration. Similarly, the relative abundance of *Haemophilus* had a significantly positive correlation with esophageal mucosal abnormalities as assessed by the EREFS (endoscopic reference score) and increasing histopathologic severity as assessed by the EoEHSS (eosinophilic esophagitis histologic scoring system). Neither microbial richness nor alpha diversity significantly correlated with the increasing density of eosinophilic infiltration, EREFS, or EoEHSS. The EoEHSS correlated with four broad salivary microbial communities representing clades (piece of a phylogeny that includes an ancestral lineage and all the descendants of that ancestor [23]) in the clustering of the taxonomic abundance profiles. These clusters tended to have differential quantitative populations of specific bacteria. For example, the cluster associated with highest EoEHSS score was defined by a higher abundance of *Haemophilus*, *Streptococcus*, *Corynebacterium*, *Moraxella*, and *Dolosigranulum*. PPIs and use of inhaled and swallowed corticosteroids were not significantly associated with differential abundance of any taxa or with changes in the microbial richness or alpha or beta diversity.

Notably, not all studies demonstrate remarkable differences in the microbiota of control and EoE patients. In a study by Laserna-Mendieta et al. [24], when compared with controls, only differences were found in beta but not in alpha analysis. In a study by Johnson et al., of adult patients with EoE using biopsy specimens [25], principal coordinate analysis ordination failed to distinguish cases from controls. There were no microbiome differences within EoE cases based on clinical phenotype, presence of atopy or endoscopic features. This study raises the question as to whether pediatric and adult EoE patients may differ in their esophageal microbiome. This study also demonstrated some differences with PPI therapy.

Data summary and pathophysiologic significance

The goals of studying the microbiome in EoE are sequential. The first goal is to document whether there are changes in the altered microbiome relative to the normal state and to disease activity. The second goal is to move beyond the associations and determine whether the changes in microbial population represent a marker of the disease or a pathophysiologic link to disease occurrence. The studies above have made great strides in achieving these goals. Within the remarkable normal diversity of the esophageal microbiota in the normal state, some consistent signals have been identified. For example, there appear in most studies to be differences in the microbiome in patients with EoE. These could be differences in abundance of bacteria and the types and communities of bacteria present. These changes do not appear in most studies to be affected by diet, proton pump inhibitor, or steroid therapy. The most consistent is the presence of *Haemophilus* (spp) in esophageal and salivary microbiome. Another difference occurs when comparing active to inactive EoE. These changes are not specifically consistent but could be in esophageal microbial abundance. A third though inconsistent finding was that patients with EoE have a decreased amount of *Actinomyces* and *Streptococcus* normally identified robustly in the esophagus of normal volunteers. A fourth finding is that there is modest but not precise correlation between salivary/mouth and esophageal microbiome not only in EoE but as has been found in normal controls [26]. This provides some support for using salivary/mouth samples as a marker for esophageal populations. Finally, an increased presence of *Haemophilus* (spp) appears consistently in these studies. This could have physiologic significance as *Haemophilus* has been implicated in other TH2 type diseases. For example, asthma [27] and chronic rhinosinusitis [28,29] are associated with increased *Haemophilus* in the local mucosal microbiome. A recent review has highlighted that there is considerable variation in the techniques for collection and analysis of the esophageal microbiome. Further characterization focusing on metabolomics and tissue biochemistry are underway [30].

Conclusions

As the role of the microbiome continues to evolve in human diseases particularly those related to immunity and allergy, its part in the pathophysiology of EoE is gaining strength. Early signals include different types, communities and abundances of microbial populations and a predominance or relative deficiency of specific bacterial phyla and/or genera. Some of these changes have been more strongly implicated in other type TH2 diseases lending support for a similar proposed role in EoE. These proposed changes include disturbance of epithelial barrier function and activation of TSLP and TLR as an integral pathway to allergic inflammation.

Future directions

- More information is needed on whether LPS specifically activated TLRs in EoE.
- The mechanisms of how esophageal dysbiosis interact with food antigens to generate the inflammatory and fibrotic response in EoE will need clarity.
- The variation in collection and analysis techniques highlight the need for further need for biomic relationship to metagenomic, transcriptional effects in the esophagus in EoE.
- The ultimate question, as in all microbiome research, is whether restoration of the esophageal microbiome to normal will become a therapeutic target in EoE. More specifically, prebiotics, probiotics, and antibiotics have been used that change the gut microbiome with some beneficial effect in some diseases.
- Studying whether these therapies will be useful for treating esophageal dysbiosis and associated disease will be an exciting direction for esophageal diseases, particularly EoE.

COI

Consulting for Celgene, Regeneron, Takeda.

References

[1] M. Brusilovsky, R. Bao, M. Rochman, A.M. Kemter, C.R. Nagler, M.E. Rothenberg, Host-microbiota interactions in the esophagus during homeostasis and allergic inflammation, Gastroenterology 162 (2) (2022) 521–534.e8. https://doi.org/10.1053/j.gastro.2021.10.002.

[2] S.M. D'Souza, K. Houston, L. Keenan, B.S. Yoo, P.J. Parekh, D.A. Johnson, Role of microbial dysbiosis in the pathogenesis of esophageal mucosal disease: A paradigm shift from acid to bacteria? World J. Gastroenterol. 27 (18) (2021) 2054–2072, https://doi.org/10.3748/wjg.v27.i18.2054. 34025064. PMCID: PMC8117736.
[3] Z. Pei, E.J. Bini, L. Yang, M. Zhou, F. Francois, M.J. Blaser, Bacterial biota in the human distal esophagus, Proc. Natl. Acad. Sci. U. S. A. 101 (12) (2004) 4250–4255. https://doi.org/10.1073/pnas.0306398101.
[4] E. Norder Grusell, G. Dahlén, M. Ruth, et al., Bacterial flora of the human oral cavity, and the upper and lower esophagus, Dis. Esophagus 26 (1) (2013) 84–90, https://doi.org/10.1111/j.1442-2050.2012.01328.x. Epub 2012 Mar 6.
[5] D.A. Katzka, K. Ravi, D.M. Geno, et al., Endoscopic mucosal impedance measurements correlate with eosinophilia and dilation of intercellular spaces in patients with eosinophilic esophagitis, Clin. Gastroenterol. Hepatol. 13 (2015) 1242–1248. e1241 https://doi.org/10.1016/j.cgh.2014.12.032.
[6] D.A. Katzka, R. Tadi, T.C. Smyrk, et al., Effects of topical steroids on tight junction proteins and spongiosis in esophageal epithelia of patients with eosinophilic esophagitis, Clin. Gastroenterol. Hepatol. 12 (2014) 1824–1829. e1821 https://doi.org/10.1016/j.cgh.2014.02.039.
[7] L. Wu, T. Oshima, M. Li, et al., Filaggrin and tight junction proteins are crucial for IL-13-mediated esophageal barrier dysfunction, Am. J. Physiol. Gastrointest. Liver Physiol. 315 (2018) G341–G350, https://doi.org/10.1152/ajpgi.00404.2017.
[8] J.C. Masterson, K.A. Biette, J.A. Hammer, et al., Epithelial HIF-1alpha/claudin-1 axis regulates barrier dysfunction in eosinophilic esophagitis, J. Clin. Invest. 129 (2019) 3224–3235, https://doi.org/10.1172/JCI126744.
[9] J.D. Sherrill, D. Wu, Z. Djukic, et al., Desmoglein-1 regulates esophageal epithelial barrier function and immune responses in eosinophilic esophagitis, Mucosal Immunol. 7 (2014) 718–729, https://doi.org/10.1038/mi.2013.90.
[10] C. Blanchard, E.M. Stucke, K. Burwinkel, et al., Coordinate interaction between IL-13 and epithelial differentiation cluster genes in eosinophilic esophagitis, J. Immunol. 184 (2010) 4033–4041, https://doi.org/10.4049/jimmunol.0903069.
[11] S.A. Brenner, S. Zacheja, M. Schäffer, K. Feilhauer, S.C. Bischoff, A. Lorentz, Soluble CD14 is essential for lipopolysaccharide-dependent activation of human intestinal mast cells from macroscopically normal as well as Crohn's disease tissue, Immunology 143 (2014) 174–183, https://doi.org/10.1111/imm.12299.
[12] F. Borriello, R. Raffaella Iannone, S. Di Somma, et al., Lipopolysaccharide-elicited TSLPR expression enriches a functionally discrete subset of human CD14(+) CD1c(+) monocytes, J. Immunol. 198 (2017) 3426–3435, https://doi.org/10.4049/jimmunol.1601497.
[13] M. Kitajima, H.C. Lee, T. Nakayama, S.F. Ziegler, TSLP enhances the function of helper type 2 cells, Eur. J. Immunol. 41 (2011) 1862–1871, https://doi.org/10.1002/eji.201041195.
[14] J.K. Harris, R. Fang, B.D. Wagner, et al., Esophageal microbiome in eosinophilic esophagitis, PLoS One 10 (2015), e0128346, https://doi.org/10.1371/journal.pone.0128346.
[15] M.S. Desai, A.M. Seekatz, N.M. Koropatkin, et al., A dietary fiber-deprived gut microbiota degrades the colonic mucus barrier and enhances pathogen susceptibility, Cell 167 (2016) 1339–1353. e1321 https://doi.org/10.1016/j.cell.2016.10.043.
[16] F. Imhann, M.J. Bonder, A.V. Vila, et al., Proton pump inhibitors affect the gut microbiome, Gut 65 (2016) 740–748, https://doi.org/10.1136/gutjnl-2015-310376.
[17] A.G. Clooney, C.N. Bernstein, W.D. Leslie, et al., A comparison of the gut microbiome between long-term users and non-users of proton pump inhibitors, Aliment. Pharmacol. Ther. 43 (2016) 974–984, https://doi.org/10.1111/apt.13568.

[18] G. Nardone, D. Compare, A. Rocco, A microbiota-centric view of diseases of the upper gastrointestinal tract, Lancet Gastroenterol. Hepatol. 2 (2017) 298–312, https://doi.org/10.1016/S2468-1253(16)30108-X.
[19] V. Di Pilato, G. Freschi, M.N. Ringressi, L. Pallecchi, G.M. Rossolini, P. Bechi, The esophageal microbiota in health and disease, Ann. N. Y. Acad. Sci. 1381 (2016) 21–33, https://doi.org/10.1111/nyas.13127.
[20] T. Kaymak, P. Hruz, J.H. Niess, Immune system and microbiome in the esophagus: implications for understanding inflammatory diseases, FEBS J. (2021), https://doi.org/10.1111/febs.16103.
[21] G.E. Norder, G. Dahlen, M. Ruth, H. Bergquist, M. Bove, The cultivable bacterial flora of the esophagus in subjects with esophagitis, Scand. J. Gastroenterol. 53 (2018) 650–656, https://doi.org/10.1080/00365521.2018.1457712.
[22] A.J. Benitez, C. Hoffmann, A.B. Muir, et al., Inflammation-associated microbiota in pediatric eosinophilic esophagitis, Microbiome 3 (2015) 23, https://doi.org/10.1186/s40168-015-0085-6.
[23] D.A. Baum, A. Cécile, B. Larget, et al., Statistical evidence for common ancestry: application to primates, Evolution 70 (2016) 1354–1363, https://doi.org/10.1111/evo.12934.
[24] E.J. Laserna-Mendieta, J.A. Fitgerald, L. Arias-Gonzalez, et al., Esophageal microbiome in active eosinophilic esophagitis and changes induced by different therapies, Sci. Rep. 11 (2021) 7113, https://doi.org/10.1038/s41598-021-86464-z.
[25] J. Johnson, E.S. Dellon, A.N. McCoy, et al., Lack of association of the esophageal microbiome in adults with eosinophilic esophagitis compared with non-EoE controls, J. Gastrointestin. Liver Dis. 30 (2021) 17–24, https://doi.org/10.15403/jgld-3049.
[26] M.K. Annavajhala, M. May, G. Compres, et al., Relationship of the esophageal microbiome and tissue gene expression and links to the oral microbiome: a randomized clinical trial, Clin. Transl. Gastroenterol. 11 (2020), e00235, https://doi.org/10.14309/ctg.0000000000000235.
[27] H. Bisgaard, M.N. Hermansen, F. Buchvald, et al., Childhood asthma after bacterial colonization of the airway in neonates, N. Engl. J. Med. 357 (2007) 1487–1495, https://doi.org/10.1056/NEJMoa052632.
[28] D. Lal, P. Keim, J. Delisle, et al., Mapping and comparing bacterial microbiota in the sinonasal cavity of healthy, allergic rhinitis, and chronic rhinosinusitis subjects, Int. Forum. Allergy Rhinol. 7 (2017) 561–569, https://doi.org/10.1002/alr.21934.
[29] T. Chalermwatanachai, R. Vilchez-Vargas, G. Holtappels, et al., Chronic rhinosinusitis with nasal polyps is characterized by dysbacteriosis of the nasal microbiota, Sci. Rep. 8 (2018) 7926, https://doi.org/10.1038/s41598-018-26327-2.
[30] J.D. Busing, M. Buendia, Y. Choksi, G. Hiremath, S.R. Das, Microbiome in eosinophilic esophagitis-metagenomic, metatranscriptomic, and metabolomic changes: a systematic review, Front. Physiol. 12 (2021) 731034, https://doi.org/10.3389/fphys.2021.731034.

CHAPTER 9

Laryngopharyngeal reflux and dysbiosis

Shabnam Sarker and Michael F. Vaezi
Department of Medicine, Division of Gastroenterology, Hepatology and Nutrition, Vanderbilt University Medical Center, Nashville, TN, United States

Introduction—The gut microbiota and its influence

The microbiota of humans is an ecosystem working in synergy and homeostasis with its host. Dysbiosis occurs when the balance is disrupted sometimes with gain of microbial species and other times with loss of species. Some bacteria can be harmful, whereas others can be protective of certain diseases [1].

The gut microbiota works in synergy to provide many beneficial effects to its human host. Bacteria product-specific enzymes allow the breakdown of consumed products to more digestible forms. Short-chain fatty acids are also made from carbohydrates and yield antiinflammatory effects [2,3]. These host and microbiome interactions lead to homeostasis, whereas disarray and consequent dysbiosis can lead to disruption in normative host function. Dysbiosis can cause significant effects in cytokine expressions and a broad array of consequent related diseases, including inflammatory bowel disease (IBD), gastrointestinal infections as well as esophagitis [4].

The composition of each host's microbiome is unique to the individual. The vast array of microorganisms, which also vary in their genetic compositions, leads to significant variations in genes within the microbiome. The composition of the microbiome can be influenced significantly by the environment. Humans are exposed to microorganisms during birth and shortly after from their mothers. The majority of our gastrointestinal (GI) colonization occurs during our early developmental years, albeit subject to changing multiple extrinsic and intrinsic factors over time. This includes but are not limited to diet, environmental, demographic, and GI motility effects. [2] The microbiome diversity plays a greater role in phenotypical variance than the actual human genome [2]. Kassam et al. showed that infection by *Clostridium difficile* has a greater influence in the colon microbiome than our genes [5]. Genomic influence has a potentially broad ranging effect on both

Esophageal Disease and the Role of the Microbiome
https://doi.org/10.1016/B978-0-323-95070-1.00008-4

health and disease. The microbiome has been linked to the development of autoimmune disorders, including type I diabetes mellitus as well as IBD. Obese vs. lean individuals have different microbiome populations, and this population can be used to determine risk of obesity [6]. The microbiota also plays a role in mental and behavioral health so much so that fecal transplantation has shown improvement in autism behaviors [7].

Conversely, antibiotics can alter the gut microbiota and lead to dysbiosis. The use of antibiotics early in life has been linked to obesity later in life as well as autoimmune conditions, including IBD [8].

Mechanisms for defense against laryngopharyngeal reflux

The upper aerodigestive tract is where the respiratory tract, including pharynx and larynx, intersects with the digestive tract from mouth to esophagus [9]. It serves as a mucosal barrier, concentrated with cells that compose the immune system [9].

The first line of defense is mucus, a viscous solution coating the surfaces to prevent injury and decrease invasion of pathogens [10]. Microbicidal peptides and Immunoglobin A (IgA) and IgG contribute to a large portion of the mucus layer. The gastric tissue also has a mucus layer that prevents injury from acids and pepsin [11]. Lipids and bicarbonate help neutralize hydrochloric acid [10,11].

The epithelial layer is the first tissue layer of defense, which is located above an acellular basement membrane and tight junctions [12]. Beneath lies a network of vascular and lymphatics housing the immune modulating cells that aid in defense [9]. Mucosal contact of foreign pathogens with the mucosa has been shown to lead to inflammation and several inflammatory disorders [9]. Laryngopharyngeal reflux (LPR) is a relatively common diagnosis evaluated in otolaryngology clinic [13]. This occurs when gastric contents directly are regurgitated or are aerosolized into the laryngopharynx [9,14].

Many patients with gastroesophageal reflux disease (GERD)-induced chronic laryngitis present with symptoms such as sore throat, globus, chronic cough, hoarseness, dysphagia, apnea, or postnasal drip, among other symptoms. However, they may not present with classic GERD symptoms of heartburn and regurgitation. A 1991 study found that of 225 patients assessed with otolaryngologic disorders having suspected GERD, only 43% reported symptoms of heartburn or acid regurgitation [15]. Additional symptoms that may be related to reflux-related laryngitis may include frequent throat clearing, dry mouth, prolonged voice warm-up time (greater than 20 to 30 min), halitosis, excess phlegm, coated tongue, throat tickle, regurgitation of food, nocturnal cough, difficulty breathing especially at night, aspiration, laryngospasm, poorly controlled asthma, or pneumonia.

Pepsin in particular, along with acid and bile salts, has been linked to mucosal injury to the laryngopharyngeal mucosa [16]. These enzymes are believed to cause chemical injury by causing dysbiosis and inflammation [17]. Ma et al. recently published a study measuring salivary pepsin in those with symptomatic GERD compared with normal controls [18]. They also measured esophageal exposure time and symptom scores of antacid therapy. They found that those with increased pepsin have greater symptoms as well as higher acid exposure time [18]. Also, increased pepsin was associated with increased oral microbiome biodiversity with predominance of Firmicutes ssp., Bacteroidetes ssp., Proteobacteria ssp., Actinobacteria ssp., and Fusobacteria ssp. consistent with findings of above studies. The increased oral microbiome diversity can have implications in upper esophageal, aerodigestive, and extra-esophageal symptoms, including LPR.

Diagnosis of LPR

LPR is an inflammatory process, associated with multiple conditions, including otitis media, rhinosinusitis, and vocal fold lesions [17]. Symptoms can range from pain, cough, dysphagia, and dysphonia (Fig. 1) [16]. During visualization, erythema in oropharynx is noted as well as posterior

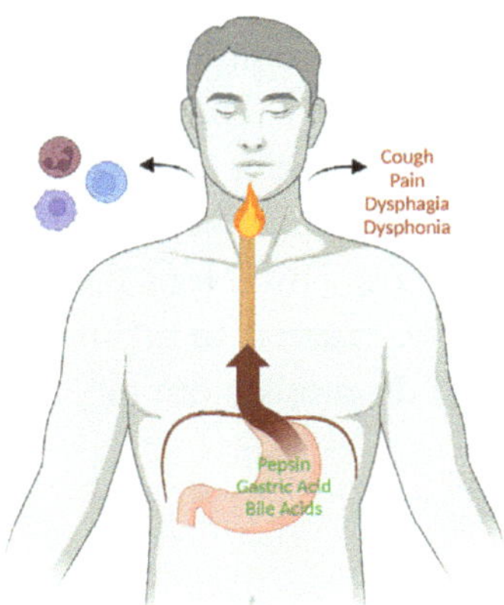

Fig. 1 LPR with reflux of gastric contents leading to inflammation and symptoms. *(Created with BioRender.com.)*

commissure hypertrophy [16]. Direct visualization using video laryngological examination performed by experienced laryngologist is important to evaluate for visual cues to mucosal injury. Laryngeal signs identified during laryngoscopy seem to be poorly specific for identifying gastroesophageal reflux. One study showed that lesions of the vocal fold may represent more specific signs for LPR, exhibiting 91% specificity and 88% response to treatment with proton pump inhibitors (PPIs) [19]. The nonspecificity of laryngoscopy may also be due to poor inter- and intraobserver variability. A randomized prospective analysis by five otolaryngologists blinded to patient information of 120 video segments of rigid fiberoptic laryngoscopy found poor inter-rater reliabilities of the laryngoscopic findings associated with LPR (intraclass correlation coefficient of 0.265), and intra-rater reliability was extremely variable for the various physical findings (Kendall correlation coefficients ranging from −0.121 to 0.837). Taken together, these studies suggest that accurate assessment of laryngeal involvement with LPR is quite difficult as interpretation of physical findings is subjective and varies among physicians. Laryngoscopic findings of erythema is highly subjective but more objective signs such as vocal cord abnormalities may be more specific for reflux-induced laryngitis. Fiberoptic evaluation using air to simulate the laryngopharynx can be used to evaluate for hypersensitivity [20]. Diagnosis also includes the use of multichannel intraluminal mucosal impedance-pH to identify acid and nonacid reflux events [16]. In addition, recent data with this device suggests that the findings from impedance-pH does not correlate with symptoms because patients can have normal findings with significant symptoms [21]. The 2008 American Gastroenterological Association (AGA) Technical Review on the management of GERD suggested that the role of pH or impedance-pH monitoring in diagnosing extraesophageal reflux is controversial and unproven. This evidence-based technical review concludes that the value of a negative pH or impedance-pH study is of greater clinical utility. The detection of pepsin and gastroduodenal enzymes may be helpful to aid in diagnosis [16]. None the less, there are no standardized guidelines for diagnostic criteria. In addition, LPR can coexist with GERD making the isolated diagnosis more difficult.

Lifestyle factors

H Chen et al. compared the microbiota of patients with LPR to healthy volunteers and found that Prevotella ssp. was more common in the LPR group, whereas Fusobacterium ssp. and Porphyromonas ssp. were lower [22]. They also evaluated the diversity of microbiome in smokers who were noted to have Lactobacillus ssp. and Clostridium ssp. [22]. Alcohol decreased Streptococcus ssp. [22] Clearly, the predominant species of microbiota vary among those with LPR and healthy individuals [22]. Lifestyle factors such as smoking and alcohol consumption also change the array of microbiota [22]. Dietary intake clearly effects the microbiota configuration and thereby may have a role in the treatment of LPR [16].

Hypersensitivity

Sensitivity to mucosal injury varies among individuals as some have hypersensitivity leading to symptoms in the setting of normal testing. Mucosal hypersensitivity is associated with LPR [23]. The exact cause of mucosal hypersensitivity is not clear and may be more than just functional symptoms. Dysbiosis is thought to play a role in hypersensitivity [17]. However, more data are needed to better understand this important and emerging concept.

Acid and nonacid esophageal reflux

The role of gastric acid in gastrointestinal disorders has been a subject of great interest and intense controversy for many years. It has been associated with heartburn, the cardinal symptom of GERD. However, the definition of GERD is still debated as those without evidence of esophageal injury can have the diagnosis, whereas others with no symptoms, appropriately labeled "silent reflux," can demonstrate either histologic, with or without macroscopic esophagitis. Given that many patients with LPR may also have concomitant heartburn and esophagitis, it is important to better understand the concept of dysbiosis in this overlap group of patients.

The influence of the oral microbiome on reflux esophagitis has shown conflicting results. Wang et al. showed there was no significant difference in the oral microbiome of those with reflux esophagitis compared with healthy volunteers [24]. However, Ahn et al. showed that distal esophageal microbiota was different in those with reflux esophagitis [25]. Yang et al. evaluated the microbiome from distal esophageal biopsies and found that genus Streptococcus ssp. was predominant in normal patients, whereas patients with esophagitis and Barrett's esophagus had predominant gram-negative anaerobes and microaerophiles with significantly elevated odds ratios of 15 and 16, respectively [1]. Gram-negative bacteria produce lipopolysaccharides which causes relaxation of the lower esophageal sphincter and can contribute to gastroesophageal reflux [26]. Recognizably, at present, it is unclear whether reflux leads to gram-negative bacteria predominance or whether the bacteria leads to reflux, although there is emerging data suggesting the former pathway as the initial pathogenesis [1]. Reflux of gastric contents can also select for more acid-resistant bacteria in the esophagus [1].

Zhou et al. found that patients with nonerosive reflux disease (NERD) have more sulfate-reducing Proteobacteria ssp. and Bacteroidetes ssp. as well as hydrogen-producing Dorea ssp. associated with higher Firmicutes ssp. composition [14]. In the same study, NERD patients had less Fusobacteria ssp. and Actinobacteria ssp. [14]. Patients with NERD have the ability to repair or prevent esophageal mucosal damage, whereas those with reflux esophagitis do not. Notably, these patients also have similar microbiota composition to control patients, suggesting the different microbiota between NERD and erosive GERD may play a role in esophageal injury [14].

Esophageal and gastric adenocarcinoma

Esophageal adenocarcinoma (EAC) is highly rare in those with LPR symptoms alone, but in those with overlap of GERD and LPR, the distal esophageal mucosa may be at risk. The composition of the oral microbiome is also different in those with gastric cancer compared with normal controls [1]. Incidence of EAC has increased exponentially in the past several decades, which is believed to be related to environmental factors [1]. Those with EAC had increased lactic acid–producing genus, including Staphylococcus ssp., Lactobacillus ssp., Bifidobacterium ssp., and Streptococcus ssp. [14]. Dysbiosis and alterations in lactate metabolism may play a role in carcinogenesis [14]. Blackett et al. found an abundant gram-positive bacteria population in those with EAC. Zhou et al. found more Firmicutes ssp. and Proteobacteria ssp. in EAC and less Actinobacteria ssp. [14]

Microbiota and bacterial overgrowth

Patients with LPR commonly have overlap symptoms of GERD and/or dyspepsia. Dysbiosis has also been shown to cause gas and bloating symptoms [27]. Gas-bloat disorders can cause refractory reflux symptoms [27]. Haworth et al. evaluated patients referred for antireflux surgery and evaluated dysbiosis via hydrogen-positive test indicating small intestinal bacterial overgrowth (SIBO) and methane-positive test indicating intestinal methane overload. The referred patients had similar acid workup and similar findings of acid exposure and number of reflux episodes being comparable. The patients with dysbiosis were more likely to have active reflux symptoms compared with those without dysbiosis [27]. Regurgitation was also more commonly reported in patients with dysbiosis [27].

Treatment and proton pump inhibitors

The majority of patients with suspected LPR are initially treated with twice daily PPIs notably higher doses than used for typical GERD. This is used as empiric therapy. Alkaline and low-fat diet and lifestyle modifications may also help in treatment and can be used in conjunction with PPIs [16]. However, PPI therapy many not be helpful in those nonacid or biliary LPR. In this population, alginate drugs may be more helpful [16]. It is important to ensure compliance with therapy and rule out other gastrointestinal disease with similar symptoms in those failing to respond to therapy [16]. Diet and lifestyle modifications may be the first step in management, particularly in mild LPR [16]. Although PPIs are widely used in patients with suspected LPR, high-quality supporting evidence remains minimal at this time. Placebo-controlled studies assessing lansoprazole, esomeprazole, pantoprazole, and rabeprazole for two to four months duration have observed no significant difference in symptoms experienced by LPR patients on placebo as compared with PPI [28,29]. Similarly, a meta-analysis of randomized controlled trials assessing PPI use for suspected GERD-related chronic laryngitis noted no benefit of PPIs over placebo [30]. On the other hand, in support of the role of acid suppression in LPR, a more recent double-blind, placebo-controlled trial noted a significant improvement in LPR symptoms with three months of esomeprazole therapy [31]. Finally, in a randomized controlled study of patients with chief complaint of postnasal drainage by Vaezi et al. [32] found a benefit for treatment with PPIs, suggesting that PPIs may do more than just suppress reflux in this group of individuals. Thus, the search for the subgroup of patients with suspected LPR or extraesophageal reflux symptoms more likely to respond to PPI therapy continues.

The use of PPIs has been shown to change the microbiotic environment and predispose to enteric infections due to decreased acid production and ability to neutralize certain pathogens [33,34]. This allows colonization of these microorganisms starting in the stomach. The composition changes to more predominant oral and oropharyngeal microbes [33,34]. PPIs have a greater influence on gut microbiome than other common medications, including antibiotics [34].

The use of PPIs have additionally been associated with SIBO [35]. Intestinal microbial changes and consequent cytokine-related stimulation of nitric oxide and cyclooxygenase (COX)-2, which promote lower esophageal sphincter relaxation and delayed gastric emptying, respectively, can have potential effects on GERD and related consequences [36].

Conclusions

There is emerging data on the oropharyngeal and gastrointestinal microbial effects associated with LPR. The microbiome can be significantly influenced by the environment, including diet, demographic factors (including smoking and alcohol) as well as use of medications such as antibiotics and PPIs. Disruption in the balance and homeostasis can predispose to GERD-related disease, including LPR.

Future directions

A clearer understanding of the precise roles of specific microbes and alterations leading to dysbiosis and disease might lead to better understanding, diagnosis, and treatments for patients with LPR. Future studies that might lead to that clearer understanding include:

- Identification of specific microbial patterns leading to LPR.
- Characterizing the effects of acid and bile salts individually in altering the aerodigestive microbiome.
- Mechanistic evaluation of microbial alterations and the effects on the sensation of heartburn and symptoms consistent with LPR.
- Identification and evaluation of targeted mitigation strategies directed at dysbiosis in LPR

COI and disclosures

Dr. Vaezi -Advisory for Ironwood, Phathom, ISOThrive, Sanofi, Bayer Pharmaceuticals; Diversatek healthcare; Patent on Mucosal Integrity Testing; Legal consultation in litigations relating to acid suppressive agents.

Dr. Sarker has no disclosures.

References

[1] L. Yang, X. Lu, C.W. Nossa, et al., Inflammation and intestinal metaplasia of the distal esophagus are associated with alterations in the microbiome, Gastroenterology 137 (2009) 588–597.
[2] J.A. Gilbert, M.J. Blaser, J.G. Caporaso, et al., Current understanding of the human microbiome, Nat. Med. 24 (2018) 392–400.
[3] P.D. Cani, M. Van Hul, C. Lefort, et al., Microbial regulation of organismal energy homeostasis, Nat. Metab. 1 (2019) 34–46.
[4] J.R. Marchesi, D.H. Adams, F. Fava, et al., The gut microbiota and host health: a new clinical frontier, Gut 65 (2016) 330–339.
[5] Z. Kassam, C.H. Lee, Y. Yuan, et al., Fecal microbiota transplantation for Clostridium difficile infection: systematic review and meta-analysis, Am. J. Gastroenterol. 108 (2013) 500–508.
[6] D. Knights, L.W. Parfrey, J. Zaneveld, et al., Human-associated microbial signatures: examining their predictive value, Cell Host Microbe 10 (2011) 292–296.
[7] D.W. Kang, J.B. Adams, A.C. Gregory, et al., Microbiota transfer therapy alters gut ecosystem and improves gastrointestinal and autism symptoms: an open-label study, Microbiome 5 (2017) 10.
[8] L. Trasande, J. Blustein, M. Liu, et al., Infant antibiotic exposures and early-life body mass, Int. J. Obes. (Lond) 37 (2013) 16–23.
[9] D. Liu, T. Qian, S. Sun, et al., Laryngopharyngeal reflux and inflammatory responses in mucosal barrier dysfunction of the upper aerodigestive tract, J. Inflamm. Res. 13 (2020) 1291–1304.
[10] F. Taherali, F. Varum, A.W. Basit, A slippery slope: on the origin, role and physiology of mucus, Adv. Drug Deliv. Rev. 124 (2018) 16–33.
[11] Y. Niv, R. Fass, The role of mucin in GERD and its complications, Nat. Rev. Gastroenterol. Hepatol. 9 (2011) 55–59.
[12] Buckley A., Turner J.R., Cell biology of tight junction barrier regulation and mucosal disease. Cold Spring Harb. Perspect. Biol. 2018;Jan; 10(1): a029314.
[13] J.R. Lechien, S. Saussez, A. Nacci, et al., Association between laryngopharyngeal reflux and benign vocal folds lesions: a systematic review, Laryngoscope 129 (2019) E329–E341.
[14] J. Zhou, P. Shrestha, Z. Qiu, et al., Distinct microbiota dysbiosis in patients with non-erosive reflux disease and esophageal adenocarcinoma, J. Clin. Med. 9 (7) (2020) 2162.
[15] J.A. Koufman, The otolaryngologic manifestations of gastroesophageal reflux disease (GERD): a clinical investigation of 225 patients using ambulatory 24-hour pH monitoring and an experimental investigation of the role of acid and pepsin in the development of laryngeal injury, Laryngoscope 101 (1991) 1–78.
[16] J.R. Lechien, L.M. Akst, A.L. Hamdan, et al., Evaluation and management of laryngopharyngeal reflux disease: state of the art review, Otolaryngol. Head Neck Surg. 160 (2019) 762–782.
[17] J.R. Lechien, N. De Vos, A. Everard, et al., Laryngopharyngeal reflux: the microbiota theory, Med. Hypotheses 146 (2021), 110460.
[18] S.D. Ma, C. Gucionne, K. Linnemeyer, et al., Oral microbiome diversity between patients with normal versus elevated salivary pepsin levels and its implication in gastroesophageal reflux disease, in: (Abstract Su 1209) Presented at Digestive Diseases Week. San Diego, CA, 2022.
[19] W. Park, D.M. Hicks, F. Khandwala, et al., Laryngopharyngeal reflux: prospective cohort study evaluating optimal dose of proton-pump inhibitor therapy and pretherapy predictors of response, Laryngoscope 115 (2005) 1230–1238.

[20] B. Famokunwa, E.S. Walsted, J.H. Hull, Assessing laryngeal function and hypersensitivity, Pulm. Pharmacol. Ther. 56 (2019) 108–115.
[21] M. Duricek, P. Banovcin, T. Halickova, et al., Acidic pharyngeal reflux does not correlate with symptoms and laryngeal injury attributed to laryngopharyngeal reflux, Dig. Dis. Sci. 64 (2019) 1270–1280.
[22] H. Chen, H. Wang, F. Yang, et al., Distinct microbiota dysbiosis in patients with laryngopharynx reflux disease compared to healthy controls, Eur. Arch. Otorhinolaryngol. (2022).
[23] J.H. Hull, A. Menon, Laryngeal hypersensitivity in chronic cough, Pulm. Pharmacol. Ther. 35 (2015) 111–116.
[24] B. Wang, Y. Zhang, Q. Zhao, et al., Patients with reflux esophagitis possess a possible different oral microbiota compared with healthy controls, Front. Pharmacol. 11 (2020) 1000.
[25] J. Ahn, C.Y. Chen, R.B. Hayes, Oral microbiome and oral and gastrointestinal cancer risk, Cancer Causes Control 23 (3) (2012) 399–404.
[26] Y.P. Fan, S. Chakder, F. Gao, et al., Inducible and neuronal nitric oxide synthase involvement in lipopolysaccharide-induced sphincteric dysfunction, Am. J. Physiol. Gastrointest. Liver Physiol. 280 (2001) G32–G42.
[27] J.J. Haworth, N. Boyle, A. Vales, et al., The prevalence of intestinal dysbiosis in patients referred for antireflux surgery, Surg. Endosc. 35 (2021) 7112–7119.
[28] T. Havas, S. Huang, M. Levy, et al., Posterior pharyngolaryngitis: double-blind randomised placebo-controlled trial of proton pump inhibitor therapy, Aust. J. Otolaryngol. 3 (1999) 243–246.
[29] J.M. Wo, J. Koopman, S.P. Harrell, et al., Double-blind, placebo-controlled trial with single-dose pantoprazole for laryngopharyngeal reflux, Am. J. Gastroenterol. 101 (2006) 1972–1978.
[30] M.A. Qadeer, C.O. Phillips, A.R. Lopez, et al., Proton pump inhibitor therapy for suspected GERD-related chronic laryngitis: a meta-analysis of randomized controlled trials, Am. J. Gastroenterol. 101 (2006) 2646–2654.
[31] O. Reichel, H. Dressel, K. Wiederänders, et al., Double-blind, placebo-controlled trial with esomeprazole for symptoms and signs associated with laryngopharyngeal reflux, Otolaryngol. Head Neck Surg. 139 (2008) 414–420.
[32] M.F. Vaezi, D.D. Hagaman, J.C. Slaughter, et al., Proton pump inhibitor therapy improves symptoms in postnasal drainage, Gastroenterology 139 (2010) 1887–1893.e1.
[33] M.A. Jackson, J.K. Goodrich, M.E. Maxan, et al., Proton pump inhibitors alter the composition of the gut microbiota, Gut 65 (2016) 749–756.
[34] F. Imhann, M.J. Bonder, A. Vich Vila, et al., Proton pump inhibitors affect the gut microbiome, Gut 65 (2016) 740–748.
[35] L. Lombardo, M. Foti, O. Ruggia, et al., Increased incidence of small intestinal bacterial overgrowth during proton pump inhibitor therapy, Clin. Gastroenterol. Hepatol. 8 (2010) 504–508.
[36] S.M. D'Souza, L.B. Cundra, B.S. Yoo, P.J. Parekh, D.A. Johnson, Microbiome and gastroesophageal disease: pathogenesis and implications for therapy, Ann. Clin. Gastroenterol. Hepatol. 4 (2020) 20–33.

CHAPTER 10

Esophageal dysbiosis and esophageal motility disorders

Amit Patel[a] and C. Prakash Gyawali[b]
[a]Division of Gastroenterology, Duke University School of Medicine and the Durham VA Medical Center, Durham, NC, United States
[b]Division of Gastroenterology, Washington University School of Medicine, Saint Louis, MO, United States

Introduction

The esophageal microbiome is a dynamic and rapidly evolving area of investigation, both in health and in disease. Accordingly, there may be potentially profound implications for diagnostic investigation, clinical care, and population health [1,2]. In contrast to the intestinal microbiome, there remain significant limitations in understanding of the esophageal microbiome and pathophysiological implications of dysbiosis (Fig. 1). In addition, relationships between the esophageal microbiome and esophageal motility disorders have not yet been methodically investigated and determined. Much of our understanding of these relationships are limited to dysbiosis in distinct esophageal processes such as gastroesophageal reflux disease (GERD), esophageal cancer, and eosinophilic esophagitis (EoE) [3]. In this chapter, we review what is known about the esophageal microbiome in the context of esophageal motility disorders and put forth speculations regarding future research opportunities and avenues within this realm.

Achalasia

Achalasia spectrum disorders are cardinal esophageal motility disorders characterized by incomplete relaxation of the lower esophageal sphincter (LES) with abnormal esophageal body peristalsis, that present with dysphagia to solids and/or liquids, bland regurgitation, retrosternal discomfort, and/or weight loss [4–6]. Limited case reports have postulated potential associations between bacterial infections and achalasia spectrum disorders, such as cardiac tuberculosis in the setting of human immunodeficiency virus [7] and *Mycobacterium goodii* pulmonary infection [8], although evidence to support direct cause-and-effect relationships for these pathogens are limited.

Esophageal Disease and the Role of the Microbiome
https://doi.org/10.1016/B978-0-323-95070-1.00004-7

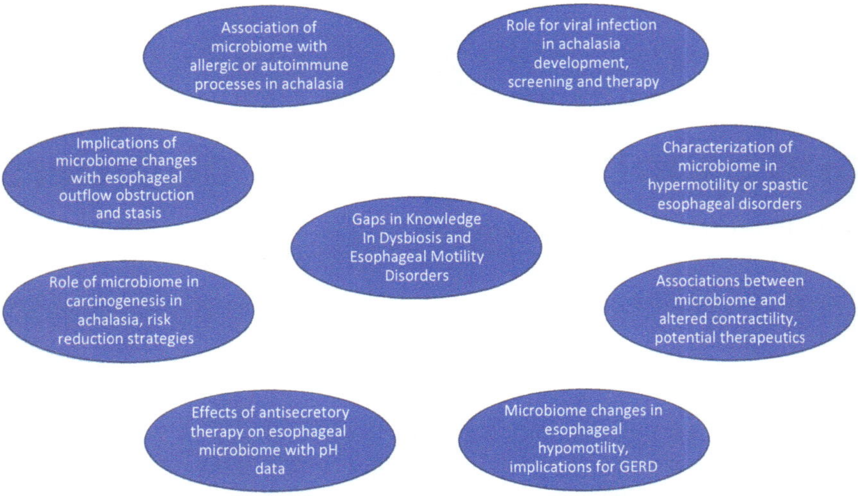

Fig. 1 Gaps in our understanding of the relationship between esophageal dysbiosis and esophageal motility disorders.

Multiple lines of investigation have suggested a potential association between viral infection and achalasia. Viral DNA has been demonstrated in esophageal tissue, and viral antibody has been measured in the serum of patients with achalasia [9,10], particularly viruses that have a predilection for squamous epithelium or neural tissue. These may include herpes simplex virus 1 (HSV-1), varicella zoster virus (VZV), and human papillomavirus (HPV) [9,11]. An European epidemiological survey with genotype–phenotype analysis included 696 idiopathic achalasia patients and 410 controls (as well as first-degree relatives of each group) [12]. This analysis demonstrated higher proportions with viral infection (most prominently for VZV infections) before onset of achalasia, suggesting a potential role for viral infections in the development of achalasia. Furthermore, the authors also found increased proportions of allergic (51% vs 40%) and (additional) autoimmune (21% vs 13%) disorders in achalasia, most prominently for psoriasis and Sjogren's syndrome, suggesting potential relationships with achalasia spectrum disorders.

In addition, an American study of 15 patients with achalasia who underwent surgical myotomy showed that 12/15 (80%) had detectable salivary VZV DNA and 13/15 (87%) had transcripts encoding VZV late gene products in surgically resected esophageal tissue [13]. These findings suggest that VZV reactivation in esophageal neurons may lead to chronic VZV infection that, in turn, could potentially affect regulation of esophageal motor function. However, given the low likelihood of reactivation solely in esophageal neurons in the setting of the massive numbers of human ganglia that can harbor latent VZV, additional factors (such as genetic considerations) are likely involved in achalasia development. Together, these investigations suggest the potential for a pathophysiologic model for the development of achalasia wherein viral infection or reactivation may prompt autoimmune responses in genetically predisposed patients that could lead to achalasia spectrum disorders (and possibly other immune-mediated processes). The potential impacts of such a model clinically cannot be understated for potential screening and management of achalasia spectrum disorders. However, these existing pathophysiologic paradigms involve systemic infection with the culprit viruses rather than resident microbiota within the esophagus. Nonetheless, the putative involvement of microbes in the pathophysiology of profound esophageal disorders like achalasia suggests that the potential exists for the esophageal microbiome to influence the development and manifestations of these motility disorders.

There are limited data that specifically address the esophageal luminal microbiome in achalasia. One single-center Japanese study compared the oral microbiome (collected by brushing the buccal mucosa with a sterile swab) and esophageal microbiome (collected by brushing the mid-esophagus during upper endoscopy, using a sheath to minimize contamination) from six patients with confirmed achalasia, who were sampled both before and after treatment with per-oral endoscopic myotomy (POEM) [14]. The investigators found that although the oral and esophageal microbiota compositions were significantly different, *Streptococcus* species were most abundant, in a more profound fashion in the esophagus compared with the oral cavity. Furthermore, after POEM, despite the absence of significant changes in the overall structure of the oral or esophageal microbiota, there was a relative abundance of *Hemophilus* and *Neisseria* species in the esophageal microbiota compared with before POEM. These findings suggest that the esophageal milieu changes after management of the esophageal outflow obstruction in achalasia from one that resembles the oral microbiome to a gram-negative predominant microbiome, potentially related to relief of the outflow obstruction and exposure to the gastric and intestinal microbial environment. The clinical implications of these changes are not clear and deserve further study. Beyond these investigations, data on the associations between achalasia spectrum disorders and the esophageal microbiome are limited at this juncture [1,15].

Esophageal motility disorders can lead to impaired esophageal clearance and stasis, potentially leading to overgrowth and/or dysbiosis. Fungal infections such as esophageal candidiasis can develop in the setting of disorders associated with stasis, such as achalasia, that potentially allow for colonization of the esophageal mucosa [16]. In fact, an analysis of 218 patients at the Mayo Clinic diagnosed with candida esophagitis found that among those 111 patients with an underlying/secondary cause for dysphagia, 15 (13.5%) had an esophageal motility disorder [17]. In addition to esophageal stasis, reduced mucosal immunity may contribute to candida growth, and the esophageal microbiome in achalasia may contribute to this. To date, however, neither the implications of overall changes in the esophageal microbiome nor the relationship between the esophageal microbiome and candida risk in achalasia have been systematically studied.

Chagas disease and megaesophagus

An infectious organism confirmed to have a causative role in gastrointestinal dysmotility is *Trypanosoma cruzi*, which is associated with Chagas disease, also known as American trypanosomiasis. Esophageal motor derangements are well described in the chronic phase of *T. cruzi* infection that can resemble aspects of achalasia [18]. Most commonly found in South America, Central America, and Mexico, untreated Chagas disease can lead to marked dilation of the esophagus (megaesophagus). There are a few studies evaluating the esophageal microbiome in megaesophagus from Chagas disease, which may mirror changes potentially at play in achalasia, because achalasia is also associated with esophageal dilation and stasis.

In a Brazilian study comparing the esophageal microbiota between 15 patients with megaesophagus associated with Chagas disease and 10 controls without esophageal disease, higher proportions of positive cultures (93% vs 40%) were seen in the megaesophagus group. There was greater variability in bacteria and predominant aerobic gram-positive bacteria, including *Streptococcus* ssp., which were the most common of this subset, and anaerobic bacteria, of which *Veillonella* ssp. were the most predominant [19]. Furthermore, in this cohort, more elevated bacterial concentrations were generally found among more profoundly dilated megaesophagus patients. In contrast, other data from Brazil suggested that while the incidence of microorganisms was high in the setting of megaesophagus, that there did not appear to be significant associations between the positivity of esophageal cultures and the degree of esophageal dilation among their cohort of patients with chagasic megaesophagus [20]. These findings, similar to the limited data available in achalasia patients, suggest that the oral microbiome is augmented and potentially replicated when there is esophageal stasis, but the degree of stasis does not seem to alter the overall microbiota.

Further Brazilian work compared saliva samples and esophageal stasis liquid samples from 15 patients with nonadvanced chagasic megaesophagus and 15 controls without esophageal disease to evaluate the potential reduction of nitrate into nitrite with the formation of volatile nitrosamines. They found that both groups had similar nitrate and nitrite concentrations in saliva and nitrate concentrations in the esophagus, but that the megaesophagus group had significantly higher esophageal nitrite and nitrate-reducing bacteria concentrations [21]. Furthermore, volatile nitrosamines were found in three of the megaesophagus patients but none of the controls, although this did not reach statistical significance. The implications of excess luminal and mucosal nitrite and nitrosamine compounds in the pathophysiology of disease progression remains unclear, because nitric oxide is a predominant neurotransmitter involved in esophageal inhibitory neurotransmission in the esophageal neural network. Thus, further in vitro and in vivo study of the effects of microbiome products on esophageal motor transmission is needed.

Squamous cell carcinoma (SCC) in the setting of achalasia

Although the nuances of the relationships between the microbiome and esophageal squamous cell carcinoma (SCC) are discussed in an earlier chapter, these associations have direct implications for achalasia and megaesophagus, where SCC can develop during long-term follow-up. An Italian cohort of 566 patients with achalasia followed prospectively for a mean of >15 years after diagnosis identified 20 who developed esophageal cancer. Of these, 15 (75%) were SCC, for an annual incidence rate of 0.18%, significantly higher than for the general population (SIR 126.9) [22]. A meta-analysis of 40 studies evaluating achalasia as a risk factor for esophageal SCC reported an incidence of SCC of 312.3 per 100,000 patient-years at risk, and an absolute risk increase for SCC of 308.1 per 100,000 patients per year [23]. A Brazilian report of 593 patients with esophageal carcinoma identified 32 patients with Chagas disease; those with locally advanced carcinoma suffered from a dismal prognosis (median overall survival of 9.1 months), reinforcing the need to better our surveillance and management approaches in this context [24]. Thus, achalasia and Chagas disease are clearly associated with an increased risk of esophageal carcinoma; the known esophageal microbiome changes in the esophageal lumen in achalasia spectrum disorders could inform and impact this risk.

Among potential mechanisms that may contribute to SCC in the setting of achalasia spectrum disorders and/or esophageal dilation, esophageal stasis may contribute to overgrowth of nitrate-reducing bacteria, leading to the production of volatile N-nitrosamines in the esophageal lumen and thus nitrosative stress, causing DNA damage in carcinogenesis [25]. For example, in the setting of increased grades of esophageal dilation among a cohort of chagasic megaesophagus, Brazilian work found more elevated concentrations of bacteria that included microorganisms capable of reducing nitrates into nitrites (*Staphylococcus, Corynebacterium, Peptostreptococcus, Veillonella* ssp), which could play a role in carcinogenesis [26]. Interestingly, the Japanese analysis discussed earlier also compared the oral and esophageal microbiota of six achalasia patients with 14 nonachalasia superficial esophageal SCC patients, showing no significant differences among oral or esophageal microbiota at the phyla or genus levels [14]. Therefore, potential cause-and-effect links of esophageal SCC to the microbiome milieu in the dilated esophagus with stasis in achalasia and Chagas disease deserve further study.

Esophageal hypermotility disorders

Pathophysiologic mechanisms involving abnormal esophageal inhibitory nerve function and imbalance between esophageal excitation and inhibition prevalent in achalasia could extend to the pathophysiology of esophageal hypermotility disorders [27]. Available data indicates that hypercontractile esophagus and type 3 achalasia can share altered neural control characterized by abnormal inhibition [28], and that hypercontractile esophagus or distal esophageal spasm may develop or progress into frank achalasia over time [29]. In this setting, some of the microbial dynamics previously discussed for achalasia may be prevalent in these hypermotility disorders, including potential contribution of viral infections to pathogenesis. An extension of this could include esophageal microbiome changes associated with esophageal stasis. Furthermore, as discussed below, there could be microbiome-related contributions to pathophysiology, or effects of microbiome products on the generation of abnormal contraction patterns. There are, however, very limited data on the esophageal microbiome in hypermotility or spastic motor disorders. Clearly, future research endeavors should focus on studying these potential relationships.

Furthermore, the esophageal microbiota could potentially be associated with alterations in contractility, drawing from data elsewhere in the gastrointestinal tract. For instance, among patients who had undergone surgery for severe chronic constipation, spontaneous contractions and carbachol/potassium-induced contractions of descending colon segment circular and longitudinal muscle strips were of higher amplitude than that seen in controls without constipation, indicating that the neuromuscular mechanisms of contraction were not just retained, but were exaggerated in some instances [30]. The authors speculated that some of these contractility changes in severe chronic constipation could have been the result of local luminal and mucosal factors, including the microbiome milieu. Interestingly, the prevalence of common probiotic genera was not reduced, suggesting against the efficacy of conventional probiotics to restore the gut microbiome in patients with constipation. Other studies have demonstrated that luminal microbiota may contribute to contractility and affect motility through alterations in microbiome metabolites, and therefore, manipulation of the microbiome could potentially impact contractility and gut motility [31].

Esophageal hypomotility disorders

There will likely be significant overlap between microbiome changes and dysbiosis between esophageal hypomotility disorders and the GERD spectrum (including Barrett's esophagus), because hypomotility can be associated with reflux disease [32,33]. This is discussed in greater detail in the earlier chapters on esophageal dysbiosis and the pathogenesis of GERD and related disorders [34]. There are differences in esophageal biofilm composition in Barrett's esophagus and GERD, with increased *Campylobacter concisus* on cultural and molecular analysis, which may be associated with increased cytokine expression [35]. Furthermore, beyond the associations between obesity and GERD [36], it has been proposed that exposure to unhealthy diets with associated obesity may change the esophageal microbiome in ways similar to the intestinal microbiome (such as increases in short-chain fatty acid producers and bile acid biosynthesis) and in this manner potentially contribute to esophageal adenocarcinoma [37]. However, these changes have not been investigated within patients with esophageal hypomotility.

From an investigative perspective, a novel Texan ex vivo model was recently proposed to mimic the esophagus and facilitate mechanistic investigations of the esophageal microbiome and resultant transcriptome, with a focus on Barrett's esophagus in the human distal esophagus [38]. This initial work suggests a distinct microbiome in Barrett's esophagus, with the potential to create cell lines that can be used for further investigations to better understand the mechanisms of microbial interactions. This line of investigation has potential to uncover esophageal microbiome changes within hypomotility disorders and potentially be extended to study other esophageal disorders.

Proton pump inhibitors (PPIs) are frequently used to manage patients with esophageal symptoms, which can also affect the gastrointestinal microbiome, and potentially the esophageal microbiome as well. Given the mechanism of action of PPIs to reduce gastric acid production, it would be expected that PPI effects on the esophageal microbiome could be more focused in areas closest to the stomach, i.e., the distal esophagus [39]. Although discussed in more comprehensive detail in the earlier chapter on acid suppression and its effects on the microbiome, recent data based on customized esophageal microbiome qPCR panel array approaches suggest that the use of PPIs can lead to a significant difference in absolute levels of actinomyces but not other organisms [40]. Furthermore, in this analysis, there were no significant associations between PPI dosing or duration of use and the absolute levels of any organism.

Multicenter clinical trial data from Korea evaluated microbiome changes and symptom response among patients with nonerosive reflux disease who were treated with 20mg esomeprazole for eight weeks [41]. The authors found that *Streptococcus, Haemophilus, Prevotella, Veillonella, Neisseria,* and *Granulicatella* ssp. were prevalent both at baseline and week 8 of PPI therapy, and that microbial compositions of the oropharynx and esophagus were not significantly affected by PPI therapy in this trial. Further investigations designed to include esophageal pH data to better understand how acidity and reflux burden are associated with the esophageal microbiome are warranted.

Among children with oropharyngeal dysphagia and with suspected GERD, there was overlap between oropharyngeal, gastric, and lung microbiomes, but the degree of similarity was lowest between the oropharynx and the lungs [42]. However, among children with aspiration based on video fluoroscopic swallow studies, the lung microbiome more closely resembled the oropharyngeal rather than the gastric microbiome. These findings suggest that microbiome changes can result from motor processes and potentially from regurgitation leading to aspiration.

Future research

- Despite preliminary provocative data, the interplay between the esophageal microbiome and esophageal motility disorders have not been adequately evaluated and deserve further investigation given the potential clinical ramifications.
- Investigations incorporating the thoughtful acquisition of esophageal microbiome datasets with associated clinical and diagnostic data from esophageal motor disease populations will be key to advancing this field.
- Augmenting our understanding of specifically how the microbiome influences esophageal motor function, and how esophageal motility disorders and stasis can shape the microbiome, carry immense implications for disease prevention, diagnosis, tailored management, and the surveillance of patients for esophageal disorders, including motor disorders.

Conflicts of interest

AP: none; CPG: Medtronic, Diversatek, Ironwood, Takeda, Johnson & Johnson, Ardelyx, Dexcel Pharma, Quintiles, ISOThrive.

References

[1] A. Hasan, L.K. Hasan, B. Schnabl, et al., Microbiome of the aerodigestive tract in health and esophageal disease, Dig. Dis. Sci. 66 (2021) 12–18.

[2] V. Di Pilato, G. Freschi, M.N. Ringressi, et al., The esophageal microbiota in health and disease, Ann. N. Y. Acad. Sci. 1381 (2016) 21–33.

[3] B. Corning, A.P. Copland, J.W. Frye, The Esophageal microbiome in health and disease, Curr. Gastroenterol. Rep. 20 (2018) 39.

[4] A. Patel, S. Posner, C.P. Gyawali, Esophageal high-resolution manometry in gastroesophageal reflux disease, JAMA 320 (2018) 1279–1280.

[5] C.P. Gyawali, D.A. Carlson, J.W. Chen, et al., ACG clinical guidelines: clinical use of Esophageal physiologic testing, Am. J. Gastroenterol. 115 (2020) 1412–1428.

[6] R. Yadlapati, P.J. Kahrilas, M.R. Fox, et al., Esophageal motility disorders on high-resolution manometry: Chicago classification version 4.0(©), Neurogastroenterol. Motil. 33 (2021), e14058.

[7] A.J. Wang, L.X. Tu, C. Yu, et al., Achalasia secondary to cardial tuberculosis caused by AIDS, J. Dig. Dis. 16 (2015) 752–753.

[8] D. Martínez-González, J. Franco, D. Navarro-Ortega, et al., Achalasia and *Mycobacterium goodii* pulmonary infection, Pediatr. Infect. Dis. J. 30 (2011) 447–448.

[9] C.S. Robertson, B.A. Martin, M. Atkinson, Varicella-zoster virus DNA in the oesophageal myenteric plexus in achalasia, Gut 34 (1993) 299–302.

[10] I. Castagliuolo, P. Brun, M. Costantini, et al., Esophageal achalasia: is the herpes simplex virus really innocent? J. Gastrointest. Surg. 8 (2004) 24–30 (discussion 30).

[11] M. Facco, P. Brun, I. Baesso, et al., T cells in the myenteric plexus of achalasia patients show a skewed TCR repertoire and react to HSV-1 antigens, Am. J. Gastroenterol. 103 (2008) 1598–1609.
[12] J. Becker, S. Niebisch, A. Ricchiuto, et al., Comprehensive epidemiological and genotype-phenotype analyses in a large European sample with idiopathic achalasia, Eur. J. Gastroenterol. Hepatol. 28 (2016) 689–695.
[13] R.D. Naik, M.F. Vaezi, A.A. Gershon, et al., Association of achalasia with active aricella zoster virus infection of the esophagus, Gastroenterology 161 (2021) 719–721.e2.
[14] K. Takahashi, H. Sato, T. Mizusawa, et al., Comparison of oral and esophageal microbiota in patients with achalasia before and after peroral endoscopic myotomy, Turk. J. Gastroenterol. 32 (2021) 42–52.
[15] C.H. Park, S.K. Lee, Exploring Esophageal microbiomes in esophageal diseases: a systematic review, J. Neurogastroenterol. Motil. 26 (2020) 171–179.
[16] P. Kumar, S. Mohan, A. Verma, et al., Candida esophagitis in achalasia cardia: case report and review of literature, Saudi J. Gastroenterol. 13 (2007) 88–90.
[17] P. Hoversten, F. Otaki, D.A. Katzka, Course of esophageal candidiasis and outcomes of patients at a single center, Clin. Gastroenterol. Hepatol. 17 (2019) 200–202.e1.
[18] F.P. Vicentine, F.A. Herbella, M.E. Allaix, et al., Comparison of idiopathic achalasia and Chagas' disease esophagopathy at the light of high-resolution manometry, Dis. Esophagus 27 (2014) 128–133.
[19] D. Pajecki, B. Zilberstein, M.A. dos Santos, et al., Megaesophagus microbiota: a qualitative and quantitative analysis, J. Gastrointest. Surg. 6 (2002) 723–729.
[20] C. Eduardo, A.B. Madureira, V.G. Lima, et al., Microflora in chagasic megaesophagus, Rev. Soc. Bras. Med. Trop. 35 (2002) 39–42.
[21] D. Pajecki, B. Zilberstein, I. Cecconello, et al., Larger amounts of nitrite and nitrate-reducing bacteria in megaesophagus of Chagas' disease than in controls, J. Gastrointest. Surg. 11 (2007) 199–203.
[22] R.M. Zagari, G. Marasco, V. Tassi, et al., Risk of squamous cell carcinoma and adenocarcinoma of the esophagus in patients with achalasia: a long-term prospectivecohort study in Italy, Am. J. Gastroenterol. 116 (2021) 289–295.
[23] F. Tustumi, W.M. Bernardo, J.R.M. da Rocha, et al., Esophageal achalasia: a risk factor for carcinoma. A systematic review and meta-analysis, Dis. Esophagus 30 (2017) 1–8.
[24] M.C.L. Martins, D.L. Miyazaki, C.C.T. Gabiatti, et al., Chagasic megaesophagus-associated carcinoma: clinical pattern and outcomes, J. Glob. Oncol. 5 (2019) 1–5.
[25] F. Tustumi, J.H.B. de Sousa, N.M. Dornelas, G.M. Rosa, M. Steinman, E.T. Bianchi, The mechanisms for the association of cancer and esophageal dysmotility disorders, Med. Sci. (Basel) 9 (2) (2021) 32, https://doi.org/10.3390/medsci9020032.
[26] D. Pajecki, B. Zilberstein, M.A. dos Santos, et al., Megaesophagus microbiota and carcinogenesis, Arq. Gastroenterol. 40 (2003) 16–19.
[27] C.P. Gyawali, Achalasia: new perspectives on an old disease, Neurogastroenterol. Motil. 28 (2016) 4–11.
[28] F. Quader, A. Mauro, E. Savarino, et al., Jackhammer esophagus with and without esophagogastric junction outflow obstruction demonstrates altered neural control resembling type 3 achalasia, Neurogastroenterol. Motil. 31 (2019), e13678.
[29] L. Huang, M. Pimentel, A. Rezaie, Do jackhammer contractions lead to achalasia? A longitudinal study, Neurogastroenterol. Motil. 29 (2017).
[30] D.R. Yarullina, M.U. Shafigullin, K.A. Sakulin, et al., Characterization of gut contractility and microbiota in patients with severe chronic constipation, PLoS One 15 (2020), e0235985.
[31] X. Ge, W. Zhao, C. Ding, et al., Potential role of fecal microbiota from patients with slow transit constipation in the regulation of gastrointestinal motility, Sci. Rep. 7 (2017) 441.

[32] C.P. Gyawali, S. Roman, A.J. Bredenoord, et al., Classification of esophageal motor findings in gastro-esophageal reflux disease: conclusions from an international consensus group, Neurogastroenterol. Motil. 29 (12) (2017), https://doi.org/10.1111/nmo.13104.
[33] S. Garbarino, A. Horton, A. Patel, The utility of esophageal motility testing in gastroesophageal reflux disease (GERD), Curr. Gastroenterol. Rep. 21 (2019) 37.
[34] E.J. Snider, D.E. Freedberg, J.A. Abrams, Potential role of the microbiome in Barrett's esophagus and esophageal adenocarcinoma, Dig. Dis. Sci. 61 (2016) 2217–2225.
[35] K.L. Blackett, S.S. Siddhi, S. Cleary, et al., Oesophageal bacterial biofilm changes in gastro-oesophageal reflux disease, Barrett's and oesophageal carcinoma: association or causality? Aliment. Pharmacol. Ther. 37 (2013) 1084–1092.
[36] B.D. Rogers, A. Patel, D. Wang, et al., Higher esophageal symptom burden in obese subjects results from increased esophageal acid exposure and not from dysmotility, Clin. Gastroenterol. Hepatol. 18 (2020) 1719–1726.
[37] N.O. Kaakoush, M.J. Morris, The oesophageal microbiome: an unexplored link in obesity-associated oesophageal adenocarcinoma, FEMS Microbiol. Ecol. 92 (10) (2016) fiw161, https://doi.org/10.1093/femsec/fiw161. Epub 2016 Jul 26 27465078.
[38] S. Cass, C. Hamilton, A. Miller, D. Jupiter, K. Khanipov, A. Booth, R. Pyles, T. Krill, G. Reep, I. Okereke, Novel ex vivo model to examine the mechanism and relationship of esophageal microbiota and disease, Biomedicine 9 (2) (2021) 142, https://doi.org/10.3390/biomedicines9020142.
[39] J. Bilello, I. Okereke, Impact of environmental and pharmacologic changes on the upper gastrointestinal microbiome, Biomedicine 9 (6) (2021) 617, https://doi.org/10.3390/biomedicines9060617.
[40] S. Tasnim, A.L. Miller, D.C. Jupiter, et al., Effects of proton pump inhibitor use on the esophageal microbial community, BMC Gastroenterol. 20 (2020) 312.
[41] C.H. Park, S.I. Seo, J.S. Kim, et al., Treatment of non-erosive reflux disease and dynamics of the esophageal microbiome: a prospective multicenter study, Sci. Rep. 10 (2020) 15154.
[42] C. Duvallet, K. Larson, S. Snapper, et al., Aerodigestive sampling reveals altered microbial exchange between lung, oropharyngeal, and gastric microbiomes in children with impaired swallow function, PLoS One 14 (2019), e0216453.

CHAPTER 11

Gastrointestinal microbiome: Evaluation of testing technologies

Igor Spivak[a,b] and Eran Elinav[a,c]
[a]Systems Immunology Department, Weizmann Institute of Science, Rehovot, Israel
[b]Medical Clinic III, University Hospital Aachen, Aachen, Germany
[c]Microbiome & Cancer Research, DKFZ, Heidelberg, Germany

Bacteria in the foregut: From assumed sterility to the exploration of complex microbial communities

The foregut has long been considered devoid of a resident microbiome. The overturn of this dogma in the 20th century has driven the identification of *Helicobacter pylori* causing destructive inflammatory processes in the foregut. This enabled the development of a targeted treatment to prevent potentially deadly complications such as ulcer hemorrhages or cancer. The current challenges in microbiome research reach from identifying microbial communities at sites where they seem to be uncommon, as in the esophagus, to deciphering the impact of whole microbial composition and function on host physiology in health and disease. An overview of advantages, limitations, and future directions in both culture-dependent and -independent approaches to identify and characterize bacteria can be found in Table 1.

The first description of bacteria in the stomach dates back to the late 19th century: Oppler [1] and Turck [2] independently found rod-shaped and lactic acid-producing bacteria in the stomachs of deceased gastric cancer patients. The bacteria first described as the Boas-Oppler bacilli were annotated to the family of *Lactobacilli* and later were found to colonize the gastric mucosa after acid suppression independently of cancer [3]. Cancer and ulcer surgery provided the first samples for bacterial culture in the early 20th century with *Staphyloccci*, *Streptococci*, and *Bacilli* identified in a minor part of the sample population [4]. These hurdles and the failure to identify single causative agents for widespread diseases led to a decreasing interest in microbial populations of the foregut, peaking in the re-application of the term "sterile" to this site until the discovery of *Helicobacter* in 1982 [5]. In 1983, the Australian pathologist J.R. Warren was able to identify a rod-shaped bacterium associated with inflammation of the mucosa in gastric biopsy specimens

Esophageal Disease and the Role of the Microbiome
https://doi.org/10.1016/B978-0-323-95070-1.00002-3

Table 1 Advantages, disadvantages and future directions of culture-based and nucleic-acid based detection methods for studies of the gastrointestinal microbiome.

Detection techniques	Culture-based	Nucleic-acid based
Advantages	– Possibility to expand small numbers of bacteria in extended culture conditions – Not sensitive to incorporated host material – Less prone to contamination during the workflow – In vitro functional validation of properties predicted through nucleic-acid based methods – Low costs	– Incorporates organisms that are hard to cultivate – Identification down to strain variant level – Provides whole community perspective – Metagenomes allow functional predictions
Limitations	– Lacks to represent whole, interconnected communities – Sampling site contamination[a]	– DNA isolation efficiency differs across strains – High host DNA background[a] – Environmental and workflow contaminants[a] – Incomplete reference databases
Future directions	– High throughput scaling coupled with modern-era detection techniques (culturomics)	– Integration into multiomics analysis pipelines – Bioinformatic advancements to identify and omit contaminants[a]

[a]Special consideration is required in low-biomass environments like the esophagus.

through a special silver staining. B. Marshall managed to isolate the germ and proposed it as an individual strain [6], allowing Goodwin et al. [7] to name this species *H. pylori* after extensive in vitro morphological and functional characterization.

The progressive development of endoscopy enabled the collection of different kinds of samples from multiple sites without the need for surgery. That also led to the kickoff to investigate the microbiota of the esophagus.

The first attempts only led to no or very low numbers of isolated and cultured bacteria, especially in healthy subjects and washes from this site (around 10^1 CFU/mL), and, consequently, small numbers of shared taxa throughout samples [8,9]. Studies in which microbial community compositions were compared in parallel liquid sampling of the stomach, esophagus, and oral cavity in gastric cancer subjects found differences in residing bacteria in-between locations [10]. These approaches were bolstered when culture-free methods capable of capturing complex community compositions began to emerge at the end of the 20th century. The target of interest became the 16S rRNA gene. It allowed to amplify DNA signatures from formerly uncharacterized bacteria, define phylogenetic distances, and identify the taxa through sequencing of the DNA [11]. In 1992, Relman et al. [12] were able to amplify the 16s rRNA gene of a bacterium found in biopsy specimens of patients suffering from Whipple's disease with a sequence coverage of >90% for the first time. This led to the phylogenetic classification of the previously uncharacterized and uncultivable causative agent further on termed *Tropheryma whipplei*. In the early 2000s, the same technology enabled a leap from identifying single bacteria in the stomach to the characterization of whole microbial communities. Broad-range polymerase chain reaction (PCR) and cloning of amplification products [13] made it possible to explore the diversity of multiple bacterial strains in the stomachs of *H. pylori* colonized patients. The constant advance in the field of sequencing technology did not stop at that point. High-throughput 454-pyrosequencing of the 16s rRNA gene with increased depth and coverage also enabled simultaneous processing of a high number of samples through barcoding, allowing to describe bacterial communities in the stomach below the genus level [14]. Although Roche's pyrosequencing started the revolution of next-generation sequencing, a technology developed by their competitors, Illumina, enabled a chemistry and chip design that facilitated the throughput while increasing cost-efficacy [15]. This way, untargeted metagenomic sequencing of short reads of the whole DNA in a sample followed by annotation to reference genome databases opened the door to identifying subclusters and functional characteristics within complex microbial communities. As this method does not rely on the amplification of bacterial target genes, studies can encompass members of the microbiome other than bacteria, such as viruses, protozoa, or fungi. Nevertheless, the low microbial biomass and the high load of eukaryote DNA especially present in esophageal samples up to now made it difficult to achieve the goals of deciphering an interconnected community of different microbes and their functional relationship [16]. Results from a small number of exploratory projects in that field are currently available as preprints [17,18]. Therefore, more data are needed to evaluate the insight added by this method to improve the understanding of the microbial communities in this part of the body. The future metagenomic workflows in exploring low biomass sites like the esophagus may benefit from recent advancements, promising an increased yield of input bacteria [19] and removal of contaminant material [20]. Driven by the limitations of genomic approaches in studying bacteria, some groups reworked culture-based methods to serve the requirements of modern-day microbiome studies. Cultivation techniques allowing the growth even of low abundant and fastidious bacteria coupled with high-throughput identification methods have defined the term "culturomics." In a low biomass environment, it opens up new opportunities for capturing the taxonomic and functional complexity [21]. Moreover, deciphering the bacterial transcriptional, metabolic and proteomic landscape in multiomics pipelines has the potential to improve the functional classification of communities and disentangle their impact on host physiology.

Culture-dependent identification of microbial communities

Bacterial culture was one of the first methods to describe the human microbiome. Nevertheless, it has long been considered outdated, especially in the light of the holistic characterization of microbial communities promised by high-throughput nucleic acid-dependent methods. Interestingly, especially in settings where the microbial biomass is considered low, as in the esophagus, culture-dependent approaches have undergone a rediscovery accompanied by refined culture conditions and supported by state-of-the-art technologies for the detection and functional characterization of microbes. This way, a veteran in the field of microbiology techniques has been termed "culturomics" to serve as a complementary tool to and compete with other omics-disciplines in the discovery of microbial patterns. One of the main expectations of this technology is to overcome the difficulty in cultivating fastidious microorganisms by providing an environment similar to their natural conditions [22]. The most important factors that need to be taken into account in the cultivation process are time, atmospheric conditions, nutrient availability, and selective inhibitors of growth.

Time: The discovery of the most prominent pathogenic microbe in the foregut has resulted from the unintentionally long cultivation of biopsy specimens from patients suffering from gastritis and peptic ulcer disease. Left in the incubator over the Easter holidays, the growth of rod-shaped bacteria was detected after 4 days, later identified as *H. pylori* [23]. *Campylobacter* constitutes another strain present in biopsies from subjects with carcinoma of the esophagus requiring a cultivation time of at least 3 days [24,25].

Atmospheric conditions: Bacteria found in the foregut originating from the oral cavity (species like *Moraxella*, *Haemophilus*, or *Streptococcus*) [26] and *Campylobacter* strains prefer either aerophilic or microaerophilic conditions, with the presence of oxygen in (sub-)atmospheric concentrations. The culture of *Helicobacter* is known to require reduced oxygen concentration, usually not exceeding 5% [27]. Other presumed colonizers of the upper GI tract, such as *Bacteroides* and *Clostridiales*, require anaerobic conditions for culture [28]. Especially strictly aerobic bacteria, sensitive to even low concentrations of O_2, require meticulous handling conditions. The currently most common method includes an anaerobic jar with an H_2 and CO_2 generating pouch for transporting the sample to an anaerobic chamber, where organisms are handled and incubated, preventing the exposure to oxygen [29].

Nutrient availability: Fastidious microorganisms often require media enriched with nutrients, for which the introduction of blood from horses or sheep into agar has become an efficient and cost-effective standard. This has become general practice in maintaining *Helicobacter* in culture [30] required for antibiotic resistance testing. Other examples that enabled the culture of bacteria in the upper GI tract include succinate, nitrate, formate, and fumarate (*Campylobacter*) [24] or sodium-acetate (*Lactobacilli*) [31]. Especially the challenging culture of strict anaerobes requires the addition of supplements like vitamins or trace elements to the media [32].

Selective inhibition: To capture the complexity of a microbial community, a broad range of culture conditions has to include the inhibition or depletion of certain members to select for others. The addition of antibiotics to the growth media is often done. As an example, nalidixic acid and colimycin are commonly added to media to select for the growth of Gram-positive bacteria. Other antimicrobial substances for selection are bile salts (Gram-negatives) or sodium-azide (Gram-positives). More specific approaches can select for certain groups of microbes such as novobiocin (*Salmonella*) [33] or ethanol for enriching spore-forming microorganisms, such as *Clostridia* [34].

In culturomics, the culturing of bacteria in various conditions is typically followed by matrix-assisted laser desorption/ionization-time of flight (MALDI-TOF) or sequencing of the 16s rRNA gene to identify the present microbial strains. The mass spectrometry (MS)-based MALDI-TOF does not require any steps of isolation or purification of microbes but allows the direct application of the cultured colonies on a target plate, where they are crystallized. The plate is introduced into the MS where the samples are bombarded by a laser, leading to the vaporization of ionized molecules. These are accelerated through an electric field onto a time TOF MS-phase, in which their time of arrival corresponds to their molecular mass and charge, generating a mass spectrum of peaks along a time axis [35]. A mapping of the mass spectra to databases allows the identification of strains reaching an accuracy level very close to the gold standard of biochemical identification [36]. The ability of this method to identify single organisms within 1 hour is of high importance for rapid decision-making in the clinical setting [37]. Further benefits of this method are relatively low costs and the potential for high-throughput use in computational pipelines to identify functional characteristics of identified microbes such as strain virulence [38], tissue invasiveness [39], or antidrug resistance [40]. Although only a few studies have taken on this technique to characterize the foregut microbiome, in a comparative approach throughout the gastrointestinal tract (GI), Mailhe et al. [41] were able to identify more than 100 species from six human individuals in aspirates from the stomach and duodenum. Compared with direct sequencing of the 16s rRNA gene, the culturomics approach resulted in a fivefold increase of detected species, whereas the overlap in-between methods did not go beyond 12%. This underlines the necessity for broader studies on a high number of subjects comparing different isolation and identification methods of resident bacteria combined with a meticulous account for contaminants on the way to characterize a common core microbiome of the foregut.

All in all, culture-dependent methods of characterizing the microbiome are of persistent value in the current era of genomics-based microbiome studies, particularly due to their potential to overcome limitations of low microbe to host DNA ratios in sample specimens from the esophagus.

Culture-independent methods of studying the microbiome

The development and refinement of microbial genome-based technologies enabled to omit the in vitro growth of viable bacteria. This led to a leap forward in the characterization of complex microbial communities as the underrepresentation of fastidious or uncultivable members vanished. On the other hand, especially in the setting of the low biomass in the esophagus, the incorporation of host DNA and contaminant microbes or their components into the workflow introduces new challenges to this promising methodology. From sampling over DNA isolation to the computational workup, appropriate measures have to be considered to produce reliable and reproducible results.

Sampling and nucleic acid isolation

Biopsies, the most common samples from the esophagus, contain a high ratio of host-to-microbe DNA. Brush sampling enabled a 10-fold increase of bacterial DNA yield with 100-fold lower levels of host DNA in this organ [42]. The stomach and duodenum allow the aspiration of fluid contents to examine microbiotas. However, the captured communities of microorganisms differ depending on the sampling method [43]. The probable reasons for this mismatch include varying abundance levels of microbes in different layers of the GI tract and incorporated host-DNA skewing the annotation of microbial DNA reads.

After sampling, the DNA must be isolated from microbes to make it accessible for further analysis. In a low biomass setting, this step may introduce several sources of variation and contamination. In most cases, the separation of nucleic acids from the microbial organisms and their purification is done with the help of all-in-one kits from commercial manufacturers. They contain standardized chemicals and matrices to lyse cells and capture isolated DNA. The primary source of variation here emerges from the efficacy of cell lysis needed to release nucleic acids [44,45]. Enzymatic, mechanical, or thermal disruption can be more or less successful in different microbes depending on the composition of their cell wall or membrane [46]. This can overemphasize the presence of certain members of the ecosystem while leaving others underrepresented.

Capturing the microbial genome

The gene encoding the small subunit of the prokaryotic ribosome contains regions highly conserved throughout species in the domains of bacteria and archaea [47]. These allow the application of several sets of "universal" primers that allow amplification from a wide range of microorganisms through PCR. The subsequent identification on next-generation sequencing platforms produces a large amount of short sequence reads in a fast and cost-effective manner. The highly variable regions in the 16s rRNA gene enable the identification of microorganisms through an alignment to phylogeny databases [48]. This bioinformatic approach can be facilitated through several software pipelines adapted for the analysis of high-throughput sequencing data such as mothur or QIIME [49,50]. This way, the relative composition of annotated or unknown taxa in a sample can be revealed. Moreover, other measures of the community structure such as richness, evenness, or diversity can be calculated. These are used to identify single bacteria or whole community phylotypes associated with health or disease. Also, they can reflect circumstances and selective pressure in the environment of the communities [51]. Nevertheless, certain limitations of this method must be taken into account, especially considering the low bacterial abundance in the esophagus: The common protocols include the amplification and sequencing of only one or two hypervariable regions of the 16s rRNA gene [52]. The primers applied at this step may insufficiently match to the amplified target gene in certain taxa, leading to an overrepresentation of others [53]. Issues with contamination through live bacteria or their DNA can emerge at every step of the workflow [54], underlining the need to introduce a sufficient number of blank control samples. The bacterial DNA signatures found within these should be reported and there must be a robust strategy to exclude contaminant taxa from being annotated as true members of the microbiota one aims to characterize [55].

Several groups have developed optimized methods that have substantially increased the sensitivity of sequencing the 16s rRNA gene. This led to pioneering insight into the microbial ecosystems in several types of solid tumors. One study focused on extending the amplified region of the 16s rRNA gene from short hypervariable regions to large parts of the whole gene [55]. In another study, increased recovery of the PCR product was achieved by linking biotin to the overhang of an amplification primer with a subsequent streptavidin bead-based pulldown [56]. Both approaches included a sophisticated bioinformatic pipeline for contamination correction. All this underlines the enormous potential these technologies harbor to decipher the role of microbial communities in previously neglected sites like the esophagus.

With advances in NGS technologies, the approach of deciphering random genomic fragments of the microbiota termed "metagenomic shotgun sequencing" emerged. In terms of taxonomic profiling, especially the computational construction of contigs and further of whole microbial genomes from sequenced fragments (metagenome-assembled genomes, MAGs) allowed an increase of the resolution down to strain-level variants [57]. This allowed the identification of subsets within microbial clades that were associated with risks or benefits in a number of diseases linked to the microbiome [58]. More importantly, this technology allows functional profiling of microbial communities present in an ecosystem. This is performed through mapping of individual reads to functional annotation databases like Kyoto Encyclopedia of Genes and Genomes (KEGG) [59] or UniProt [60]. Assembled contigs can also be searched for protein-coding sequences [61]. The vast number of features in metagenomic data allows to reconstruct pathways for a comprehensible integration of the function of single members or the complete microbial community. Pathway databases such as KEGG [62] or MetaCyc [63] and whole bioinformatic pipelines such as MetaPathways [64] or the Human Microbiome Project Unified Metabolic Analysis Network (HUMAnN) [65] enable the exploration of microbial functional resources. This way, a number of enriched pathways associated with diseases of the esophagus were identified in the to date only published metagenomic dataset from this organ [66].

Recognizably, there are certain limitations to this approach. Deep taxonomic and functional characterization requires comprehensible reference databases of complete genomes. The availability of these from cultured bacteria or MAGs is limited, especially from microbiotas that are hard to culture or found at low abundances [67]. Samples from sites with low microbial biomass, as found in the esophagus, are especially challenging for metagenomic workflows. First, sequencing pipelines require a minimal amount of input DNA, which can surpass the yield from organs with low bacterial colonization and incorporation of high amounts of host DNA [68]. The latter issue emerged in metagenomic shotgun sequencing of the esophageal microbiome [66]. Second, the high sensitivity of this method to detect microbial DNA fragments and the output of relative abundance can heavily skew the results toward reagent and environmental contaminants. This has led to certain cases of misleading interpretation for residing microbiomes [69]. Recent advances to overcome the problem of low microbial DNA input into sequencing pipelines include the amplification of small fragments of microbial DNA digested through restriction enzymes [19]. Moreover, chemical tagging of motifs highly abundant in bacterial DNA with subsequent computational filtering has been proposed to remove contaminant reads introduced from the environment, examiners, and reagents [20]. Furthermore, the bacterial metagenomic abundance does not necessarily allow conclusions about their activity, as they can persist in a quiescent state with little to no gene expression [70]. To account for this issue, the future incorporation of the microbial metatranscriptome, metaproteome, metabolome, and epigenome into multiomics pipelines [61] can disentangle the true role microorganisms play in certain sites of the body in health and disease. To date, no such characterization of the esophagus has been undertaken, although approaches from other low-biomass environments like the lung [71] have demonstrated feasibility.

Conclusions

The new focus in microbiome research on interconnected microbial communities and functional characteristics has not spared the esophagus and other organs of the foregut. Methods both culture-based and -independent provided the first insight into what could be the bacterial taxa residing in this part of the digestive tract. A significant challenge for the reliability of microbial detection pipelines is the low microbial biomass, especially prevalent in the esophagus, concomitant with bias through high amounts of incorporated host nucleic acids. Recent advances in pipelines based on culture and sequencing of bacterial genomes have the potential to overcome these boundaries and enable insight into microbial ecosystems associated with a multitude of diseases in this part of the body.

Future directions

- Characterizing resident microbiomes based on genomic approaches can provide insight into complex patterns of microbial functions and interactions with the host in the esophagus. Measures to address the issues associated with the low microbial abundance in this organ are being trialed and will hopefully contribute to reliable and reproducible future workflows.
- Advanced culture-dependent approaches ("culturomics") can augment to handle low microbial biomasses and reduce distortions introduced through host nucleic acids.
- In the future, multiomics platforms disentangling the microbial transcriptional, proteomic, metabolomic, and epigenomic landscape might also contribute to a holistic understanding of the impact of the microbiome on host physiological processes in the esophagus.

Conflict of interest disclosures

Eran Elinav E.E. is a scientific founder of DayTwo and BiomX, and a paid consultant to Hello Inside and Aposense.

References

[1] B. Oppler, Zur Kenntniss des Mageninhalts beim Carcinoma ventriculi1, Dtsch. Med. Wochenschr. (1895).

[2] F.B. Turck, The early diagnosis of carcinoma of the stomach, with the bacteriology of the stomach contents, JAMA XXIV (1895) 317–319, https://doi.org/10.1001/jama.1895.02430090019001f.

[3] P.G. Heinemann, E.E. Ecker, A study of the Boas-Oppler Bacillus, J. Bacteriol. 1 (1916) 435–444, https://doi.org/10.1128/jb.1.4.435-444.1916.

[4] M. Barber, R.H. Franklin, Bacteriology of stomach and duodenum in cases of peptic ulcer and gastric carcinoma, Br. Med. J. 1 (1946) 951–953.

[5] M. Rajilic-Stojanovic, C. Figueiredo, A. Smet, R. Hansen, J. Kupcinskas, T. Rokkas, et al., Systematic review: gastric microbiota in health and disease, Aliment. Pharmacol. Ther. 51 (2020) 582–602, https://doi.org/10.1111/apt.15650.

[6] J.R. Warren, B. Marshall, Unidentified curved bacilli on gastric epithelium in active chronic gastritis, Lancet 1 (1983) 1273–1275.

[7] C.S. Goodwin, J.A. Armstrong, T. Chilvers, M. Peters, M.D. Collins, L. Sly, et al., Transfer of *Campylobacter pylori* and *Campylobacter mustelae* to *Helicobacter* gen. nov. as *Helicobacter pylori* comb. nov. and *Helicobacter mustelae* comb. nov., respectively, Int. J. Syst. Bacteriol. 39 (1989) 397–405, https://doi.org/10.1099/00207713-39-4-397.

[8] D. Gagliardi, S. Makihara, P.R. Corsi, A. de Toledo Viana, M.V. Wiczer, S. Nakakubo, et al., Microbial flora of the normal esophagus, Dis. Esophagus 11 (1998) 248–250, https://doi.org/10.1093/dote/11.4.248.

[9] D. Pajecki, B. Zilberstein, M.A.A. dos Santos, J.A. Ubriaco, A.G. Quintanilha, I. Cecconello, et al., Megaesophagus microbiota: a qualitative and quantitative analysis, J. Gastrointest. Surg. 6 (2002) 723–729, https://doi.org/10.1016/s1091-255x(02)00028-8.

[10] S. Sjöstedt, L. Kager, A. Heimdahl, C.E. Nord, Microbial colonization of tumors in relation to the upper gastrointestinal tract in patients with gastric carcinoma, Ann. Surg. 207 (1988) 341–346, https://doi.org/10.1097/00000658-198803000-00020.

[11] F. Sanger, S. Nicklen, A.R. Coulson, DNA sequencing with chain-terminating inhibitors, Proc. Natl. Acad. Sci. U. S. A. 74 (1977) 5463–5467, https://doi.org/10.1073/pnas.74.12.5463.

[12] D.A. Relman, T.M. Schmidt, R.P. MacDermott, S. Falkow, Identification of the uncultured bacillus of Whipple's disease, N. Engl. J. Med. 327 (1992) 293–301, https://doi.org/10.1056/NEJM199207303270501.

[13] E.M. Bik, P.B. Eckburg, S.R. Gill, K.E. Nelson, E.A. Purdom, F. Francois, et al., Molecular analysis of the bacterial microbiota in the human stomach, Proc. Natl. Acad. Sci. U. S. A. 103 (2006) 732–737, https://doi.org/10.1073/pnas.0506655103.

[14] A.F. Andersson, M. Lindberg, H. Jakobsson, F. Bäckhed, P. Nyrén, L. Engstrand, Comparative analysis of human gut microbiota by barcoded pyrosequencing, PLoS One 3 (2008), e2836, https://doi.org/10.1371/journal.pone.0002836.

[15] C. Quince, A.W. Walker, J.T. Simpson, N.J. Loman, N. Segata, Shotgun metagenomics, from sampling to analysis, Nat. Biotechnol. 35 (2017) 833–844, https://doi.org/10.1038/nbt.3935.

[16] R. Eisenhofer, J.J. Minich, C. Marotz, A. Cooper, R. Knight, L.S. Weyrich, Contamination in low microbial biomass microbiome studies: issues and recommendations, Trends Microbiol. 27 (2019) 105–117, https://doi.org/10.1016/j.tim.2018.11.003.

[17] C. Sorini, K.P. Tripathi, S. Wu, S.M. Higdon, J. Wang, L. Cheng, et al., Metagenomic and single-cell RNA-seq survey of the *H. pylori*-infected stomach in asymptomatic individuals, BioRxiv (2021), https://doi.org/10.1101/2021.12.04.21267139.
[18] X. Huang, S. Ma, J. Niu, J. Li, X. Wang, X. Duan, et al., Microbiome insights into the perforated duodenal ulcer via metagenomic sequencing, ResearchSquare (2019), https://doi.org/10.21203/rs.2.16622/v2.
[19] Z. Sun, S. Huang, P. Zhu, L. Tzehau, H. Zhao, J. Lv, et al., Species-resolved sequencing of low-biomass or degraded microbiomes using 2bRAD-M, Genome Biol. 23 (2022) 36, https://doi.org/10.1186/s13059-021-02576-9.
[20] O. Mzava, A.P. Cheng, A. Chang, S. Smalling, L.-A.D. Kounatse, J. Lenz, et al., A metagenomic DNA sequencing assay that is robust against environmental DNA contamination, BioRxiv (2021), https://doi.org/10.1101/2021.11.22.469599.
[21] J.-C. Lagier, G. Dubourg, M. Million, F. Cadoret, M. Bilen, F. Fenollar, et al., Culturing the human microbiota and culturomics, Nat. Rev. Microbiol. 16 (2018) 540–550, https://doi.org/10.1038/s41579-018-0041-0.
[22] T. Kaeberlein, K. Lewis, S.S. Epstein, Isolating "uncultivable" microorganisms in pure culture in a simulated natural environment, Science 296 (2002) 1127–1129, https://doi.org/10.1126/science.1070633.
[23] B.J. Marshall, J.R. Warren, Unidentified curved bacilli in the stomach of patients with gastritis and peptic ulceration, Lancet 1 (1984) 1311–1315, https://doi.org/10.1016/s0140-6736(84)91816-6.
[24] K.L. Blackett, S.S. Siddhi, S. Cleary, H. Steed, M.H. Miller, S. Macfarlane, et al., Oesophageal bacterial biofilm changes in gastro-oesophageal reflux disease, Barrett's and oesophageal carcinoma: association or causality? Aliment. Pharmacol. Ther. 37 (2013) 1084–1092, https://doi.org/10.1111/apt.12317.
[25] S. Macfarlane, E. Furrie, G.T. Macfarlane, J.F. Dillon, Microbial colonization of the upper gastrointestinal tract in patients with Barrett's esophagus, Clin. Infect. Dis. 45 (2007) 29–38, https://doi.org/10.1086/518578.
[26] E. Norder Grusell, G. Dahlén, M. Ruth, L. Ny, M. Quiding-Järbrink, H. Bergquist, et al., Bacterial flora of the human oral cavity, and the upper and lower esophagus, Dis. Esophagus 26 (2013) 84–90, https://doi.org/10.1111/j.1442-2050.2012.01328.x.
[27] L. Nichols, M. Sughayer, P.C. DeGirolami, K. Balogh, D. Pleskow, K. Eichelberger, et al., Evaluation of diagnostic methods for *Helicobacter pylori* gastritis, Am. J. Clin. Pathol. 95 (1991) 769–773, https://doi.org/10.1093/ajcp/95.6.769.
[28] B. Zilberstein, A.G. Quintanilha, M.A.A. Santos, D. Pajecki, E.G. Moura, P.R.A. Alves, et al., Digestive tract microbiota in healthy volunteers, Clinics 62 (2007) 47–54, https://doi.org/10.1590/s1807-59322007000100008.
[29] P.H. Summanen, M. McTeague, M.L. Väisänen, C.A. Strong, S.M. Finegold, Comparison of recovery of anaerobic bacteria using the Anoxomat, anaerobic chamber, and GasPak jar systems, Anaerobe 5 (1999) 5–9, https://doi.org/10.1006/anae.1999.0184.
[30] T.G. Blanchard, J.G. Nedrud, Laboratory maintenance of helicobacter species, in: Current Protocols in Microbiology, 2006, https://doi.org/10.1002/9780471729259.mc08b01s00. (Chapter 8). Unit8B.1.
[31] K.L. Mason, J.R. Erb Downward, N.R. Falkowski, V.B. Young, J.Y. Kao, G.B. Huffnagle, Interplay between the gastric bacterial microbiota and *Candida albicans* during postantibiotic recolonization and gastritis, Infect. Immun. 80 (2012) 150–158, https://doi.org/10.1128/IAI.05162-11.
[32] P. Scherer, H. Sahm, Effect of trace elements and vitamins on the growth of *Methanosarcina barkeri*, Acta Biotechnol. 1 (1981) 57–65, https://doi.org/10.1002/abio.370010108.
[33] A.N. Jensen, G. Sørensen, D.L. Baggesen, R. Bødker, J. Hoorfar, Addition of Novobiocin in pre-enrichment step can improve *Salmonella* culture protocol of modified

semisolid Rappaport–Vassiliadis, J. Microbiol. Methods 55 (2003) 249–255, https://doi.org/10.1016/S0167-7012(03)00158-1.
[34] J.R. Koransky, S.D. Allen, V.R. Dowell Jr., Use of ethanol for selective isolation of sporeforming microorganisms, Appl. Environ. Microbiol. 35 (1978) 762–765, https://doi.org/10.1128/aem.35.4.762-765.1978.
[35] A. Croxatto, G. Prod'hom, G. Greub, Applications of MALDI-TOF mass spectrometry in clinical diagnostic microbiology, FEMS Microbiol. Rev. 36 (2012) 380–407, https://doi.org/10.1111/j.1574-6976.2011.00298.x.
[36] X.-F. Chen, X. Hou, M. Xiao, L. Zhang, J.-W. Cheng, M.-L. Zhou, et al., Matrix-assisted laser desorption/ionization time of flight mass spectrometry (MALDI-TOF MS) analysis for the identification of pathogenic microorganisms: a review, Microorganisms 9 (2021), https://doi.org/10.3390/microorganisms9071536.
[37] G. Nyvang Hartmeyer, A. Kvistholm Jensen, S. Böcher, M. Damkjaer Bartels, M. Pedersen, M. Engell Clausen, et al., Mass spectrometry: pneumococcal meningitis verified and *Brucella* species identified in less than half an hour, Scand. J. Infect. Dis. 42 (2010) 716–718, https://doi.org/10.3109/00365541003754493.
[38] J.-W. Cheng, C. Liu, T. Kudinha, M. Xiao, S.-Y. Yu, C.-X. Yang, et al., Use of matrix-assisted laser desorption ionization–time of flight mass spectrometry to identify MLST clade 4 *Clostridium difficile* isolates, Diagn. Microbiol. Infect. Dis. 92 (2018) 19–24, https://doi.org/10.1016/j.diagmicrobio.2018.04.011.
[39] H.-C. Lin, J.-J. Lu, L.-C. Lin, C.-M. Ho, K.-P. Hwang, Y.-C. Liu, et al., Identification of a proteomic biomarker associated with invasive ST1, serotype VI Group B *Streptococcus* by MALDI-TOF MS, J. Microbiol. Immunol. Infect. 52 (2019) 81–89, https://doi.org/10.1016/j.jmii.2017.11.007.
[40] A.F. Lau, H. Wang, R.A. Weingarten, S.K. Drake, A.F. Suffredini, M.K. Garfield, et al., A rapid matrix-assisted laser desorption ionization-time of flight mass spectrometry-based method for single-plasmid tracking in an outbreak of carbapenem-resistant Enterobacteriaceae, J. Clin. Microbiol. 52 (2014) 2804–2812, https://doi.org/10.1128/JCM.00694-14.
[41] M. Mailhe, D. Ricaboni, V. Vitton, J.-M. Gonzalez, D. Bachar, G. Dubourg, et al., Repertoire of the gut microbiota from stomach to colon using culturomics and next-generation sequencing, BMC Microbiol. 18 (2018) 157, https://doi.org/10.1186/s12866-018-1304-7.
[42] A. Gall, J. Fero, C. McCoy, B.C. Claywell, C.A. Sanchez, P.L. Blount, et al., Bacterial composition of the human upper gastrointestinal tract microbiome is dynamic and associated with genomic instability in a Barrett's esophagus cohort, PLoS One 10 (2015), e0129055, https://doi.org/10.1371/journal.pone.0129055.
[43] C. Schulz, K. Schütte, N. Koch, R. Vilchez-Vargas, M.L. Wos-Oxley, A.P.A. Oxley, et al., The active bacterial assemblages of the upper GI tract in individuals with and without *Helicobacter* infection, Gut 67 (2018) 216–225, https://doi.org/10.1136/gutjnl-2016-312904.
[44] O. Inceoglu, E.F. Hoogwout, P. Hill, J.D. van Elsas, Effect of DNA extraction method on the apparent microbial diversity of soil, Appl. Environ. Microbiol. 76 (2010) 3378–3382, https://doi.org/10.1128/AEM.02715-09.
[45] A. Salonen, J. Nikkilä, J. Jalanka-Tuovinen, O. Immonen, M. Rajilić-Stojanović, R.A. Kekkonen, et al., Comparative analysis of fecal DNA extraction methods with phylogenetic microarray: effective recovery of bacterial and archaeal DNA using mechanical cell lysis, J. Microbiol. Methods 81 (2010) 127–134, https://doi.org/10.1016/j.mimet.2010.02.007.
[46] S. Yuan, D.B. Cohen, J. Ravel, Z. Abdo, L.J. Forney, Evaluation of methods for the extraction and purification of DNA from the human microbiome, PLoS One 7 (2012), e33865, https://doi.org/10.1371/journal.pone.0033865.

[47] C.R. Woese, G.E. Fox, Phylogenetic structure of the prokaryotic domain: the primary kingdoms, Proc. Natl. Acad. Sci. U. S. A. 74 (1977) 5088–5090, https://doi.org/10.1073/pnas.74.11.5088.
[48] D. McDonald, M.N. Price, J. Goodrich, E.P. Nawrocki, T.Z. DeSantis, A. Probst, et al., An improved Greengenes taxonomy with explicit ranks for ecological and evolutionary analyses of bacteria and archaea, ISME J 6 (2012) 610–618, https://doi.org/10.1038/ismej.2011.139.
[49] P.D. Schloss, S.L. Westcott, T. Ryabin, J.R. Hall, M. Hartmann, E.B. Hollister, et al., Introducing mothur: open-source, platform-independent, community-supported software for describing and comparing microbial communities, Appl. Environ. Microbiol. 75 (2009) 7537–7541, https://doi.org/10.1128/AEM.01541-09.
[50] J.G. Caporaso, J. Kuczynski, J. Stombaugh, K. Bittinger, F.D. Bushman, E.K. Costello, et al., QIIME allows analysis of high-throughput community sequencing data, Nat. Methods 7 (2010) 335–336, https://doi.org/10.1038/nmeth.f.303.
[51] M.J. Cox, W.O.C.M. Cookson, M.F. Moffatt, Sequencing the human microbiome in health and disease, Hum. Mol. Genet. 22 (2013) R88–R94, https://doi.org/10.1093/hmg/ddt398.
[52] J.G. Caporaso, C.L. Lauber, W.A. Walters, D. Berg-Lyons, J. Huntley, N. Fierer, et al., Ultra-high-throughput microbial community analysis on the Illumina HiSeq and MiSeq platforms, ISME J. 6 (2012) 1621–1624, https://doi.org/10.1038/ismej.2012.8.
[53] A. Klindworth, E. Pruesse, T. Schweer, J. Peplies, C. Quast, M. Horn, et al., Evaluation of general 16S ribosomal RNA gene PCR primers for classical and next-generation sequencing-based diversity studies, Nucleic Acids Res. 41 (2013), e1, https://doi.org/10.1093/nar/gks808.
[54] M.C. de Goffau, S. Lager, S.J. Salter, J. Wagner, A. Kronbichler, D.S. Charnock-Jones, et al., Recognizing the reagent microbiome, Nat. Microbiol. 3 (2018) 851–853, https://doi.org/10.1038/s41564-018-0202-y.
[55] D. Nejman, I. Livyatan, G. Fuks, N. Gavert, Y. Zwang, L.T. Geller, et al., The human tumor microbiome is composed of tumor type-specific intracellular bacteria, Science 368 (2020) 973–980, https://doi.org/10.1126/science.aay9189.
[56] A. Fu, B. Yao, T. Dong, Y. Chen, J. Yao, Y. Liu, et al., Tumor-resident intracellular microbiota promotes metastatic colonization in breast cancer, Cell 185 (2022) 1356–1372.e26, https://doi.org/10.1016/j.cell.2022.02.027.
[57] D.T. Truong, A. Tett, E. Pasolli, C. Huttenhower, N. Segata, Microbial strain-level population structure and genetic diversity from metagenomes, Genome Res. 27 (2017) 626–638, https://doi.org/10.1101/gr.216242.116.
[58] T. Liwinski, A. Leshem, E. Elinav, Breakthroughs and bottlenecks in microbiome research, Trends Mol. Med. 27 (2021) 298–301, https://doi.org/10.1016/j.molmed.2021.01.003.
[59] M. Kanehisa, Y. Sato, M. Kawashima, M. Furumichi, M. Tanabe, KEGG as a reference resource for gene and protein annotation, Nucleic Acids Res. 44 (2016) D457–D462, https://doi.org/10.1093/nar/gkv1070.
[60] UniProt Consortium, UniProt: a worldwide hub of protein knowledge, Nucleic Acids Res. 47 (2019) D506–D515, https://doi.org/10.1093/nar/gky1049.
[61] E.A. Franzosa, T. Hsu, A. Sirota-Madi, A. Shafquat, G. Abu-Ali, X.C. Morgan, et al., Sequencing and beyond: integrating molecular "omics" for microbial community profiling, Nat. Rev. Microbiol. 13 (2015) 360–372, https://doi.org/10.1038/nrmicro3451.
[62] M. Kanehisa, S. Goto, Y. Sato, M. Kawashima, M. Furumichi, M. Tanabe, Data, information, knowledge and principle: back to metabolism in KEGG, Nucleic Acids Res. 42 (2014) D199–D205, https://doi.org/10.1093/nar/gkt1076.

[63] R. Caspi, T. Altman, R. Billington, K. Dreher, H. Foerster, C.A. Fulcher, et al., The MetaCyc database of metabolic pathways and enzymes and the BioCyc collection of pathway/genome databases, Nucleic Acids Res. 42 (2014) D459–D471, https://doi.org/10.1093/nar/gkt1103.
[64] K.M. Konwar, N.W. Hanson, A.P. Pagé, S.J. Hallam, MetaPathways: a modular pipeline for constructing pathway/genome databases from environmental sequence information, BMC Bioinformatics 14 (2013) 202, https://doi.org/10.1186/1471-2105-14-202.
[65] S. Abubucker, N. Segata, J. Goll, A.M. Schubert, J. Izard, B.L. Cantarel, et al., Metabolic reconstruction for metagenomic data and its application to the human microbiome, PLoS Comput. Biol. 8 (2012), e1002358, https://doi.org/10.1371/journal.pcbi.1002358.
[66] N.P. Deshpande, S.M. Riordan, N. Castaño-Rodríguez, M.R. Wilkins, N.O. Kaakoush, Signatures within the esophageal microbiome are associated with host genetics, age, and disease, Microbiome 6 (2018) 227, https://doi.org/10.1186/s40168-018-0611-4.
[67] B.S. Beresford-Jones, S.C. Forster, M.D. Stares, G. Notley, E. Viciani, H.P. Browne, et al., The mouse gastrointestinal Bacteria catalogue enables translation between the mouse and human gut microbiotas via functional mapping, Cell Host Microbe 30 (2022) 124–138.e8, https://doi.org/10.1016/j.chom.2021.12.003.
[68] M. Chafee, L. Maignien, S.L. Simmons, The effects of variable sample biomass on comparative metagenomics, Environ. Microbiol. 17 (2015) 2239–2253, https://doi.org/10.1111/1462-2920.12668.
[69] K. Aagaard, J. Ma, K.M. Antony, R. Ganu, J. Petrosino, J. Versalovic, The placenta harbors a unique microbiome, Sci. Transl. Med. 6 (2014) 237ra65, https://doi.org/10.1126/scitranslmed.3008599.
[70] M. Schirmer, E.A. Franzosa, J. Lloyd-Price, L.J. McIver, R. Schwager, T.W. Poon, et al., Dynamics of metatranscription in the inflammatory bowel disease gut microbiome, Nat. Microbiol. 3 (2018) 337–346, https://doi.org/10.1038/s41564-017-0089-z.
[71] L. Ren, R. Zhang, J. Rao, Y. Xiao, Z. Zhang, B. Yang, et al., Transcriptionally active lung microbiome and its association with bacterial biomass and host inflammatory status, MSystems 3 (2018), https://doi.org/10.1128/mSystems.00199-18.

CHAPTER 12

Esophagogastric microbiome—Internal and external influences effecting variance

Byung Soo Yoo[a], Alsiddig Elmahdi[a], Ana Vilela[a], Steve M. D'Souza[a], Kevin V. Houston[b], and David A. Johnson[c]
[a]Department of Internal Medicine, Eastern Virginia Medical School, Norfolk, VA, United States
[b]Department of Internal Medicine, Virginia Commonwealth University, Richmond, VA, United States
[c]Division of Gastroenterology, Department of Internal Medicine, Eastern Virginia Medical School, Norfolk, VA, United States

Introduction

In health, the esophagogastric microbiome is a physiologic balance with the host. Multiple internal and external factors may disrupt this balance, with a resultant dysbiosis that may cause a diseased state. It is of emerging importance thereby to comprehend the normal esophagogastric microbiome as well as potential factors influencing the biomic derangements.

Mannell et al. first described the microbiome in the normal esophagus, which comprises of *Streptococcus viridans*, *Haemophilus influenza*, *Neisseria catarrhalis*, *Streptococcus group B*, *Streptococcus faecalis*, and *Klebsiella pneumonia* [2]. Subsequently, further studies have elucidated that the majority of the subjects shared some common genera: *Streptococcus*, *Prevotella*, and *Veillonella* [1]. Filon et al. utilized Enterotest (esophageal string test) in 15 healthy individuals to sample esophageal microbiome and compared to that of traditional biopsy samples. The bacterial composition on 16S RNA gene sequencing showed similar results with *Streptococcus*, *Prevotella*, and *Veillonella* as the most common genera [1]. It is notable that these pioneer studies were later confirmed with different techniques and identified the microbiome in a normal healthy esophagus [1].

In 2009, Yang et al. suggested a classification of the esophageal microbiome into two groups: type I and type II [3]. Type I microbiome is dominated by gram-positive taxa and represents healthy esophagus. Type II microbiome is dominated by gram-negative taxa and often seen in patients with gastroesophageal reflux disease (GERD) and Barrett's esophagus (BE) [3]. The taxa in type I microbiome includes *Streptococcus* and *Firmicutes*,

Esophageal Disease and the Role of the Microbiome
https://doi.org/10.1016/B978-0-323-95070-1.00006-0

whereas type II microbiome includes *Prevotella, Veillonella, Haemophilus, Neisseria, Rothia, Granulicatella, Campylobacter, Porphyromonas, Fusobacterium,* and *Actionmyces* [3]. The study demonstrated that inflammation and intestinal metaplasia are associated with esophageal microbiome dysbiosis [3].

Owing to the advancements in both collection and sequencing tools, numerous studies investigated the esophageal microbiomes under different disease conditions. In addition, factors beyond disease status that influence the microbiome have been identified. In this chapter, we will attempt to summarize the most recent evidence for each factor (internal and external) which affects the esophagogastric microbiome.

Internal factors (Fig. 1)

Age

The gastrointestinal (GI) microbiome changes composition with time, genetics, and environmental factors [4]. The gut microbiota has been an area of active study which demonstrated colonization of microbiome from the early stages of life [4]. Yatsuneko et al. showed that phylogenetic composition markedly varies during the first 3 years of life and then stabilizes into adult-like composition [5]. It is conceivable that the esophageal microbiome has a similar shift in microbiome because of related internal and environmental factors.

Particularly, the esophageal microbiome has shown to be affected by age in its composition [6]. In adults from the age of 30 to 60, a shotgun

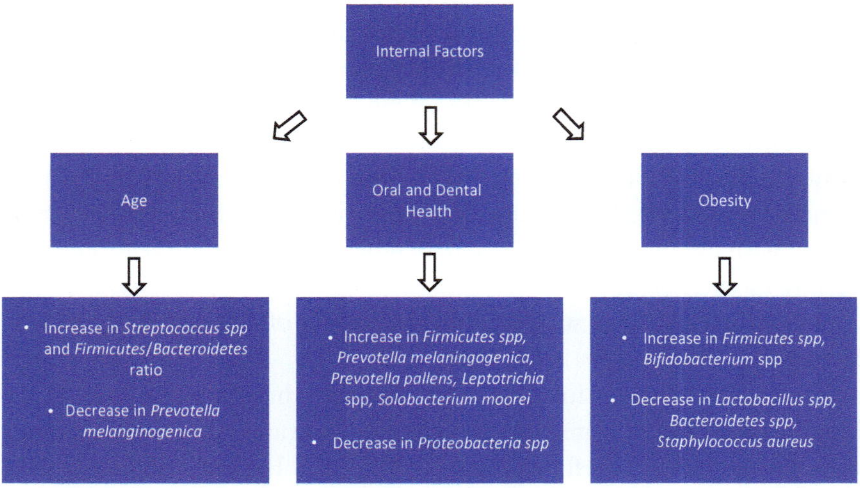

Fig. 1 Internal factors that influence esophageal microbiome.

sequencing has found that age positively correlated with *Streptococcus* spp. and inversely correlated with *Prevotella melaninogenica* [6]. Deshpande et al. also demonstrated that age significantly impacted the bacterial nucleotide biosynthesis pathways, which in turn contributes to the fluctuation of the microbiome [6].

Similar to the gut microbiome in which healthy adults have predominating *Bacteroides* and *Firmicutes* phyla and less abundant *Proteobacteria, Actinobacteria, Fusobacteria,* and *Verrucomicrobia* phyla, esophageal microbiome demonstrates a close relationship where gram positives (*Streptococcus* spp) increases with age and gram-negatives (*Prevotella* spp) decreases with time [4,6]. Notably, *Firmicutes/Bacteroidetes* (F/B) ratio increases from birth to adulthood in gut microbiome, with a predictable similar transition of composition observed in esophagus with time [4].

Oral and dental health

Oral bacteria have shown to affect the gut and esophagogastric diseases [7]. It is seen that oral and oropharyngeal microbiome travels through the GI system through saliva, food, and drinks [7]. Atarashi et al. demonstrated that there is an increasing level of oral flora in the fecal microbiota of patients with GERD, ulcerative colitis, primary sclerosing cholangitis, and alcoholism [8]. The species identified were *Rothia, Streptococcus, Neisseria, Prevotella,* and *Gamella* [8]. In addition, members of *Klebsiella* were significantly higher in abundance within the patients with Crohn's disease and alcoholism compared with that of the healthy patients [8]. In addition, Olsen et al. demonstrated that mice with induced periodontitis (with *Porphyromonas gingivalis*) showed periodontal inflammation that affects gut mechanical and immune functions [7].

The importance of oral microbial composition on both periodontal and esophageal diseases was demonstrated by Kawar et al. [9] The study observed two different groups: GERD patients using proton pump inhibitors (PPIs) and those not using PPI. In GERD patients without PPI, 17 different bacterial taxa were found to be reduced in abundance, notably *P. melaninogenica, Prevotella pallens, Leptotrichia,* and *Solobacterium moorei* [9]. In addition, the oral microbial composition of GERD patients using PPI demonstrated minimal differences compared with that of healthy subjects [9]. This study highlights the association between periodontitis and GERD, and the oral microbial composition is the key that bridges the two systems.

In addition to GERD, BE, a premalignant condition characterized by intestinal metaplasia of the esophagus, also showed a relationship with oral microbiota [10]. Snider et al. studied a total of 49 patients (32 patients with BE and 17 healthy subjects) and used 16S rRNA gene sequencing of saliva samples to compare the microbiome of the two groups. The oral microbiome of BE patients had significant increase of *Firmicutes* and decrease of *Proteobacteria* compared with that of the control [11]. The relative abundance of *Lautropia, Streptococcus*, and genus of *Bacteroidales* also signified a distinguishable point for BE against control patients. In addition, decreased abundance of *Veillonella* and increased *Enterobacteriaceae* spp. were also noted in patients with high-grade dysplasia or esophageal adenocarcinoma [11]. With these pivotal studies, oral hygiene and microbial composition offer an aspect that can influence esophageal microbiome and diseases; however, further data are necessary to fully elucidate the pathophysiology of dysbiosis and esophageal diseases.

Obesity

Obesity is a growing public health concern, and there has been a steady increase in prevalence of obesity and severe obesity [12]. Obesity is associated with a wide range of chronic medical conditions, including cardiovascular diseases, hypertension, type 2 diabetes mellitus, chronic kidney disease, and chronic liver disease [13]. This association is largely due to its pathophysiology involving chronic inflammatory state [14]. The chronic inflammatory state is well demonstrated by increases in proinflammatory cytokines [14]. The study by Hotamisligil et al. showed that in obese premenopausal women, the tumor necrosis factor (TNF)-a were expressed 2.5 times more than that of the healthy women [15]. In addition, proinflammatory cytokines such as interleukin (IL)-6, IL-1b, and chemokine (C—C Motif) ligand (CCL2) are increasingly expressed in obese patients [16,17]. NLRP3 inflammasome and toll-like receptors (TLRs) are also seen to contribute to the inflammatory state [18,19]. In association with the proinflammatory state in obesity, there is a microbial dysbiosis through the GI system.

In murine models, there is a significant decrease in the abundance of *Lactobacillus* spp. in the distal esophagus of high fat-diet fed rats [20]. Another study with similar design with high-fat diet (HFD) mice, there was decreased abundance of gram-negative *Bacteroides* and gram-positive *Clostridia* and *bifidobacterial* [21]. The human studies regarding obesity also coincide with similar dysbiosis. Ley et al. identified that in obese patients there is an increase of *Firmicutes* and decrease in *Bacteroidetes* within the fecal samples compared with that of the lean patients [22]. Similar changes are also seen in younger patients. Kalliomaki et al. studied children at birth and followed up to 7 years. The collected fecal samples showed increased *Bifidobacterium* taxa and decreased *Staphylococcus aureus* in normal children compared with overweight/obese children [23]. As seen in these studies, the dysbiosis within the GI system is observed in obesity; however, the clinical studies on esophagus are limited. One study by Lin et al. investigated 659 healthy Chinese adults with body mass index (BMI) ranging from 15.0 to 35.7 and collected upper esophageal samples. Using the 16S rRNA gene DNA microarray, the study found that an increase in BMI was associated with beta-diversity of microbial composition [24]. It is conceivable that the pathophysiology involved in obesity also plays a role in the esophagus and its microbiome and potentially contribute to chronic esophageal diseases.

External factors (Fig. 2)

Diet

Fiber and fat intake are known to modify the colonic microbiome, with reported adverse health effects. Mehta et al. found that high fiber diets resulted in reduced risk of colon cancer partially through decreased proportions of *Fusobacterium nucleatum* [25]. Furthermore, several elimination diets such as low-FODMAP (fermentable oligosaccharides, disaccharides, monosaccharides and polyols), specific carbohydrate diet, and semivegetarian have been shown to modulate disease at least partly through probiotic and prebiotic properties in the biome [26]. Although the effects of diet

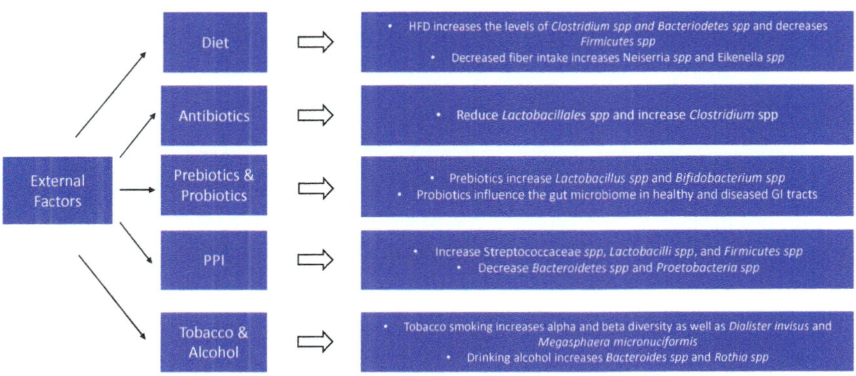

Fig. 2 External factors that influence esophageal microbiome.

on the esophageal biome are not as well understood, recent studies have shown some relationship [27].

Most of the effects of diet on the esophageal microbiome have been explored in rodent models. A 2017 study by Kaakoush et al. showed that the esophageal microbiome of rats fed a chronic HFD tended toward alterations in their biome with increased *Clostridium* throughout the biome. Interestingly, changes in *Fusobacterium*, *Rothia*, and *Granulicatella* were strongly correlated with changes in expression of obesogenic markers such as PPARA (peroxisome proliferator-activated receptor alpha) and CRAT (carnitine acetyltransferase) and carcinogenic cytokines such as IL18, indicating that cross-talk between the microbiome and host gene expression may be influenced by diet [28]. The implications of this on disease can be quite significant. Using a murine model, Stephens Much et al. showed that a HFD promoted esophageal dysplasia through alterations of the esophageal microbiome [29].

In humans, a HFD does affect the esophageal microbiome [30]. This is apparent when comparing diets of geographically distinct individuals [30]. Residents in urban areas are more likely to be obese with diets high in simple sugars, animal protein, and fat [30]. Alternatively, those in rural areas are more likely to have diets high in fiber. It appears that these communities tend to trend toward different gut and esophageal microbiome profiles [30]. Rural individuals tend to have microbiomes rich in *Succinivibrio*, *Treponema*, and *Prevotella*, organisms which support digestion of fiber. Urban individuals, however, have increased levels of *Bacteroidetes* and decreased *Firmicute*, a profile that is associated with chronic esophageal inflammation and dysbiosis [31].

In 2018, Nobel et al. showed that increasing fiber intake was associated with increased *Firmicutes* ($p = 0.04$) and decreased overall abundance of gram negative (GN) bacteria ($p = 0.03$). Decreased fiber intake was associated with increased GN bacteria such as *Neisseria* and *Eikenella*. Interestingly, the study showed no association between fat intake and microbiome composition [27].

Notably lacking at present are studies that could explore the relationship between fat consumption and the esophageal biome as well as effects of ketogenic, low carbohydrate diets, and other elimination diets. It would also be interesting to investigate the differences in effects of diet on the esophageal biome with respect to individual variations in genetics and disease. Beyond diet, it has also been suggested that bariatric surgery may alter the esophageal biome, and this effect is also worth investigating in a world where obesity is an increasing problem [30].

Antibiotics

Antibiotics have long been used to modulate disease in the GI tract. There is evidence that their use can alter the microbiome composition of the stomach and the lower GI tract toward dysbiosis [26,30]. This may be due to a variety of direct as well as indirect mechanisms. Indirectly, the depletion of specific organisms results in reduced production of metabolites which act as nutrients for other organisms [32]. In addition, indirect effects of gut transit times can affect biomic changes. Directly, broad-spectrum antibiotics have bactericidal or bacteriostatic activity that inhibit specific gut microbiota.

When it comes to the esophageal biome, effects of antibiotics have not been studied as extensively. There is data in rodent models, however, that may suggest some effect. Sawada et al. showed that treatment of postesophagojejunostomy rats with penicillin G and streptomycin resulted in reduced proportions of esophageal *Lactobacillales* and increased proportion of esophageal *Clostridium* subtypes when compared with rats not receiving antibiotics [33].

Although the implications of these changes are not well understood, it is possible that these changes may result in reduced *Lactobacillales* favorable effects. These benefits include antiinflammation, antimicrobial, and antioxidant properties. It should be noted, however, that *Lactobacillales* represent a diverse group of metabolically and physiologically distinct organisms and further studies should investigate antibiotic effects on specific subtypes [34].

In the human esophagus, some studies have shown alterations in the biome. Tasnim et al. showed that antibiotic treatment of *Helicobacter pylori* can lead to resistance which increases the colonization of resistant *H. pylori* in the distal esophagus as well as reduction of *Lactobacillus* species. Other studies have suggested that antibiotic use can increase the proportion of *Streptococcus*, the esophageal biome, and in doing so can result in a change from a Type 2 to a Type 1 biome in diseased individuals [30]. In summary, the effects of antibiotics on the human esophageal biome are conflicting and require further investigation. Although there seem to be some effect, the exact mechanism and outcome of antibiotic use on esophageal microbiome remains to be seen and placebo-controlled trials are required to draw any substantial conclusions. Furthermore, it would be interesting to compare the influence of different antibiotics such as fluoroquinolones and macrolides, whose effects on microbiome diversity have been minimally studied in the context of *H. pylori* resistance [30].

Prebiotics and probiotics

Although probiotics are defined as "live organisms, which when administered in adequate amounts, confer a health benefit on the host," [26] a prebiotic is "a substrate that is selectively utilized by host microorganisms conferring a health benefit" [35]. Both have been shown to affect the composition of the gut microbiome [36]. Most understood is the capacity of prebiotics to increase the abundance of lactic acid bacteria such as *Lactobacillus* and *Bifidobacterium*, which themselves are commonly used as probiotics [37]. Despite the abundance of studies investigating the effects of pre- and probiotics on the gut biome, there is a scarcity of evidence on changes in the esophageal biome specifically.

For a product to be considered a prebiotic, it must fulfill three criteria. It must be resistant to GI absorption, it must be fermented by intestinal microbiota, and it must selectively stimulate growth or activity of intestinal bacteria associated with improved health [37]. They are metabolized by anaerobic microbes to short-chain fatty acids (SCFA) in the gut promoting the growth of *lactobacilli* and *bifidobacteria*. The results of increased SCFA seems to be antiinflammatory in the GI tract. Unfortunately, there is a lack of substantial literature on the effects of prebiotics on the esophageal microbiome specifically.

Lactic acid probiotics such as *Lactobacillus acidophilus*, *Lactobacillus casei*, and *Bifidobacterium* species have shown a capacity to manipulate the gut microbiota in healthy as well as diseased GI tracts [38]. Other probiotics such as *Escherichia coli* Nissle 1917 have been shown to suppress the growth of pathogenic microbes and modulate the inflammatory response [26]. Interestingly, *E. coli* Nissle 1917 and *Lactobacilli* subspecies have shown a capacity to reduce production of carcinogenic compounds associated with increased risk of esophageal carcinoma [39]. Presently, there are no reports of animal models evaluating these probiotics as an antineoplastic approach.

Proton pump inhibitors

PPIs are widely used acid suppressive medications.

They are well tolerated generally, but the long-term reduction of gastric acid secretion can increase the risk of bacterial overgrowth and lead to changes in the gut microbiome [30,40].

The use of PPIs has been associated with alterations, including altered microbial pattern with decreased microflora diversity. Jackson et al.

evaluated the association between PPI use and gut microbiome using RNA amplification from fecal samples in healthy twins and found significant increase in the abundance of oral and upper GI tract commensals, particularly increase in *Streptococcaceae* and other *lactobacilli* [41]. Amir et al. demonstrated that treatment with PPIs altered the microbial communities in the gastric fluids and the esophageal tissue causing an increase in the abundance of *Firmicutes* and a decrease in the abundance of *Bacteroidetes* and *Proteobacteria* [42]. These results suggest that a more neutral environment secondary to PPI use is favorable for certain bacteria to thrive, having an effect beyond acid suppression, and therefore, supporting a possible role in altering the gastroesophageal microbiome.

Tobacco and alcohol

Cigarette smoking is the leading cause of preventable disease and death in the United States. In 2019, an estimated 50.6 million US adults reported currently using any tobacco products [43]. Cigarette smoke contains many toxic substances that may reach the GI tract and can induce microbiota dysbiosis [44]. Similarly, excessive alcohol use has been described for causing intestinal inflammation, disruption of the gut epithelial integrity, and consequently producing alterations in the GI microbiome [45,46].

There are multiple studies that investigated dysbiosis with smoking in the GI tract, but only few sought their relationship in the esophagus. A study including 278 Chinese men investigated the association between the microbiome and tobacco use [47]. Samples were obtained from esophageal mucosa through esophageal balloon cytology, demonstrating that current cigarette smokers tended to have increased alpha and beta diversity of their upper GI tract microbiota compared with former cigarette smokers and nonsmokers. The study also found *Dialister invisus* and *Megasphaera micronuciformis* species to be significantly associated with the current smoking status [47]. One possible mechanism to explain the changes in microbiota could be related to the immunosuppressive nature of tobacco. Another study found that IgG antibody levels were lower among current and former smokers than never smokers [48].

In addition, a study done in 2021 investigated the effects of smoking and drinking on the microbiota of the esophagus on its three segments (upper, middle, and lower) in healthy individuals. The esophageal mucosa was sampled using brush specimens undergoing upper endoscopy. Studied 76 patients were divided into two groups: nonsmoking/nondrinking group

(60 patients) and the smoking/drinking group (16 patients). The results revealed that there were significant differences between samples from the two groups. In the middle esophagus samples, the abundance of *Bacteroides* in the smoking/drinking group was higher than that of the nonsmoking/nondrinking group, and in the lower esophageal samples, the abundance of *Rothia* was higher in the smoking/drinking group compared with the nonsmoking/nondrinking group [49].

For smoking and alcohol consumption, there are many factors described above that may be exerting effects at the esophageal level and impact the local microbiome. Furthermore, additional basic and clinical research may help us gain more insight into the network of smoking, alcohol, and microbiome–host interactions.

Conclusions

In conclusion, studies have shown the importance of esophagogastric microbiome in esophageal disease. Understanding the factors in variances of the microbiome can pose the key to changing clinical outcomes. Although factors such as age cannot be changed as it is a natural progression of the patient, diet, medications, and social habits of smoking and drinking propose a venue for positive changes.

Future directions

- In general, genetics may play a role in resistance or susceptibility to dysbiosis in esophagus. There is limited data on genetics and esophagogastric microbiome which can explain the microbiomic variances in similar diseased populations.
- Exercise and specific lifestyles may play a role in stimulating the healthy bacteria in the esophagus. We have a limited understanding in this topic.

Author contributions

Dr. Johnson DA contributed to the construction of the project; all authors wrote and edited the manuscript.

Conflict of interest

None.

References

[1] C.H. Park, S.K. Lee, Exploring esophageal microbiomes in esophageal diseases: a systematic review, J. Neurogastroenterol. Motil. 26 (2) (2020) 171–179, https://doi.org/10.5056/JNM19240.

[2] A. Mannell, M. Plant, J. Frolich, The microflora of the oesophagus, Ann. R. Coll. Surg. Engl. 65 (3) (1983) 152.

[3] L. Yang, N. Chaudhary, J. Baghdadi, Z. Pei, Microbiome in reflux disorders and esophageal adenocarcinoma, Cancer J. 20 (3) (2014) 207–210, https://doi.org/10.1097/PPO.0000000000000044.

[4] M. Kim, B.A. Benayoun, The microbiome: an emerging key player in aging and longevity, Transl. Med. Aging 4 (2020) 103–116, https://doi.org/10.1016/j.tma.2020.07.004.

[5] T. Yatsunenko, F.E. Rey, M.J. Manary, I. Trehan, M.G. Dominguez-Bello, M. Contreras, M. Magris, G. Hidalgo, R.N. Baldassano, A.P. Anokhin, A.C. Heath, B. Warner, J. Reeder, J. Kuczynski, J.G. Caporaso, C.A. Lozupone, C. Lauber, J.C. Clemente, D. Knights, R. Knight, J.I. Gordon, Human gut microbiome viewed across age and geography, Nature 486 (7402) (2012) 222–227, https://doi.org/10.1038/nature11053.

[6] N.P. Deshpande, S.M. Riordan, N. Castaño-Rodríguez, M.R. Wilkins, N.O. Kaakoush, Signatures within the esophageal microbiome are associated with host genetics, age, and disease, Microbiome 6 (1) (2018), https://doi.org/10.1186/s40168-018-0611-4.

[7] I. Olsen, K. Yamazaki, Can oral bacteria affect the microbiome of the gut? J. Oral Microbiol. 11 (1) (2019), https://doi.org/10.1080/20002297.2019.1586422.

[8] K. Atarashi, W. Suda, C. Luo, T. Kawaguchi, I. Motoo, S. Narushima, Y. Kiguchi, K. Yasuma, E. Watanabe, T. Tanoue, C.A. Thaiss, M. Sato, K. Toyooka, H.S. Said, H. Yamagami, S.A. Rice, D. Gevers, R.C. Johnson, J.A. Segre, K. Chen, J.K. Kolls, E. Elinav, H. Morita, R.J. Xavier, M. Hattori, K. Honda, Ectopic colonization of oral bacteria in the intestine drives T H 1 cell induction and inflammation, Science (1979) 358 (6361) (2017) 359–365, https://doi.org/10.1126/SCIENCE.AAN4526.

[9] N. Kawar, S.G. Park, J.L. Schwartz, N. Callahan, A. Obrez, B. Yang, Z. Chen, G.R. Adami, Salivary microbiome with gastroesophageal reflux disease and treatment, Sci. Rep. 11 (1) (2021), https://doi.org/10.1038/S41598-020-80170-Y.

[10] Z. Zhang, G. Curran, D. Altinok Dindar, Y. Wu, H. Wu, T. Sharpton, L. Zhao, D. Lieberman, F. Otaki, Insights into the oral microbiome and Barrett's esophagus early detection: a narrative review, Clin. Transl. Gastroenterol. 12 (9) (2021), e00390, https://doi.org/10.14309/CTG.0000000000000390.

[11] E.J. Snider, G. Compres, D.E. Freedberg, M.J. Giddins, H. Khiabanian, C.J. Lightdale, Y.R. Nobel, N.C. Toussaint, A.C. Uhlemann, J.A. Abrams, Barrett's esophagus is associated with a distinct oral microbiome, Clin. Transl. Gastroenterol. 9 (3) (2018), https://doi.org/10.1038/S41424-018-0005-8.

[12] C.M. Hales, M.D. Carroll, C.D. Fryar, C.L. Ogden, Prevalence of obesity and severe obesity among adults: United States, 2017–2018, NCHS Data Brief 360 (2020) 1–8.

[13] R.K. Masters, E.N. Reither, D.A. Powers, Y.C. Yang, A.E. Burger, B.G. Link, The impact of obesity on US mortality levels: the importance of age and cohort factors in population estimates, Am. J. Public Health 103 (10) (2013) 1895–1901, https://doi.org/10.2105/AJPH.2013.301379.

[14] A.R. Saltiel, J.M. Olefsky, Inflammatory mechanisms linking obesity and metabolic disease, J. Clin. Invest. 127 (1) (2017) 1–4, https://doi.org/10.1172/JCI92035.

[15] G.S. Hotamisligil, P. Arner, J.F. Caro, R.L. Atkinson, B.M. Spiegelman, Increased adipose tissue expression of tumor necrosis factor-α in human obesity and insulin

resistance, J. Clin. Investig. 95 (5) (1995) 2409–2415, https://doi.org/10.1172/JCI117936.
[16] S.E. Shoelson, J. Lee, A.B. Goldfine, Inflammation and insulin resistance, J. Clin. Invest. 116 (7) (2006) 1793–1801, https://doi.org/10.1172/JCI29069.
[17] A.H. Berg, P.E. Scherer, Adipose tissue, inflammation, and cardiovascular disease, Circ. Res. 96 (9) (2005) 939–949, https://doi.org/10.1161/01.RES.0000163635.62927.34.
[18] K. Schroder, R. Zhou, J. Tschopp, The NLRP3 inflammasome: a sensor for metabolic danger? Science (1979) 327 (5963) (2010) 296–300, https://doi.org/10.1126/science.1184003.
[19] H. Shi, M.V. Kokoeva, K. Inouye, I. Tzameli, H. Yin, J.S. Flier, TLR4 links innate immunity and fatty acid-induced insulin resistance, J. Clin. Invest. 116 (11) (2006) 3015–3025, https://doi.org/10.1172/JCI28898.
[20] X. Zhao, X.W. Liu, N. Xie, X.H. Wang, Y. Cui, J.W. Yang, L.L. Chen, F.G. Lu, Lactobacillus species shift in distal esophagus of high-fat-diet-fed rats, World J. Gastroenterol. 17 (26) (2011) 3151–3157, https://doi.org/10.3748/WJG.V17.I26.3151.
[21] P.D. Cani, J. Amar, M.A. Iglesias, M. Poggi, C. Knauf, D. Bastelica, A.M. Neyrinck, F. Fava, K.M. Tuohy, C. Chabo, A. Waget, E. Delmée, B. Cousin, T. Sulpice, B. Chamontin, J. Ferrières, J.F. Tanti, G.R. Gibson, L. Casteilla, N.M. Delzenne, M.C. Alessi, R. Burcelin, Metabolic endotoxemia initiates obesity and insulin resistance, Diabetes 56 (7) (2007) 1761–1772, https://doi.org/10.2337/DB06-1491.
[22] R.E. Ley, Obesity and the human microbiome, Curr. Opin. Gastroenterol. 26 (1) (2010) 5–11, https://doi.org/10.1097/MOG.0B013E328333D751.
[23] M. Kalliomäki, M.C. Collado, S. Salminen, E. Isolauri, Early differences in fecal microbiota composition in children may predict overweight, Am. J. Clin. Nutr. 87 (3) (2008) 534–538, https://doi.org/10.1093/AJCN/87.3.534.
[24] S.W. Lin, N.D. Freedman, J. Shi, M.H. Gail, E. Vogtmann, G. Yu, V. Klepac-Ceraj, B.J. Paster, B.A. Dye, G.Q. Wang, W.Q. Wei, J.H. Fan, Y.L. Qiao, S.M. Dawsey, C.C. Abnet, Beta-diversity metrics of the upper digestive tract microbiome are associated with body mass index, Obesity (Silver Spring) 23 (4) (2015) 862–869, https://doi.org/10.1002/OBY.21020.
[25] R.S. Mehta, R. Nishihara, Y. Cao, M. Song, K. Mima, Z.R. Qian, J.A. Nowak, K. Kosumi, T. Hamada, Y. Masugi, S. Bullman, D.A. Drew, A.D. Kostic, T.T. Fung, W.S. Garrett, C. Huttenhower, K. Wu, J.A. Meyerhardt, X. Zhang, W.C. Willett, E.L. Giovannucci, C.S. Fuchs, A.T. Chan, S. Ogino, Association of dietary patterns with risk of colorectal cancer subtypes classified by *Fusobacterium nucleatum* in tumor tissue, JAMA Oncol. 3 (7) (2017) 921–927, https://doi.org/10.1001/JAMAONCOL.2016.6374.
[26] M. Matijašić, T. Meštrović, M. Perić, H. Paljetak, M. Panek, D.V. Bender, D.L. Kelečić, V.D. Krznarić, Modulating composition and metabolic activity of the gut microbiota in IBD patients, Int. J. Mol. Sci. 17 (4) (2016), https://doi.org/10.3390/IJMS17040578.
[27] Y.R. Nobel, E.J. Snider, G. Compres, D.E. Freedberg, H. Khiabanian, C.J. Lightdale, N.C. Toussaint, J.A. Abrams, Increasing dietary fiber intake is associated with a distinct esophageal microbiome, Clin. Transl. Gastroenterol. 9 (10) (2018) 199, https://doi.org/10.1038/S41424-018-0067-7.
[28] N.O. Kaakoush, V. Lecomte, C.A. Maloney, M.J. Morris, Cross-talk among metabolic parameters, esophageal microbiota, and host gene expression following chronic exposure to an obesogenic diet, Sci. Rep. 7 (2017), https://doi.org/10.1038/SREP45753.
[29] N.S. Münch, H.Y. Fang, J. Ingermann, H.C. Maurer, A. Anand, V. Kellner, V. Sahm, M. Wiethaler, T. Baumeister, F. Wein, H. Einwächter, F. Bolze, M. Klingenspor, D. Haller, M. Kavanagh, J. Lysaght, R. Friedman, A.J. Dannenberg, M. Pollak,

P.R. Holt, S. Muthupalani, J.G. Fox, M.T. Whary, Y. Lee, T.Y. Ren, R. Elliot, R. Fitzgerald, K. Steiger, R.M. Schmid, T.C. Wang, M. Quante, High-fat diet accelerates carcinogenesis in a mouse model of Barrett's esophagus via interleukin 8 and alterations to the gut microbiome, Gastroenterology 157 (2) (2019) 492–506.e2, https://doi.org/10.1053/J.GASTRO.2019.04.013.

[30] J. Bilello, I. Okereke, Impact of environmental and pharmacologic changes on the upper gastrointestinal microbiome, Biomedicine 9 (6) (2021), https://doi.org/10.3390/BIOMEDICINES9060617.

[31] I. Okereke, C. Hamilton, A. Wenholz, V. Jala, T. Giang, S. Reynolds, A. Miller, R. Pyles, Associations of the microbiome and esophageal disease, J. Thorac. Dis. 11 (Suppl 12) (2019) S1588, https://doi.org/10.21037/JTD.2019.05.82.

[32] S. Zhang, D.C. Chen, L.M. Chen, Facing a new challenge: the adverse effects of antibiotics on gut microbiota and host immunity, Chin Med J (Engl) 132 (10) (2019) 1135, https://doi.org/10.1097/CM9.0000000000000245.

[33] A. Sawada, Y. Fujiwara, Y. Nagami, F. Tanaka, H. Yamagami, T. Tanigawa, M. Shiba, K. Tominaga, T. Watanabe, M. Gi, H. Wanibuchi, T. Arakawa, Alteration of esophageal microbiome by antibiotic treatment does not affect incidence of rat esophageal adenocarcinoma, Dig. Dis. Sci. 61 (11) (2016) 3161–3168, https://doi.org/10.1007/S10620-016-4263-6.

[34] C. Mekadim, J. Killer, R. Pechar, J. Mrázek, Fragment of the aspartyl-tRNA synthetase applicable as a shared classification and phylogenetic marker in particular representatives of the order Lactobacillales, Folia Microbiol. (Praha) 64 (1) (2019) 113–120, https://doi.org/10.1007/S12223-018-0638-8.

[35] G.R. Gibson, R. Hutkins, M.E. Sanders, S.L. Prescott, R.A. Reimer, S.J. Salminen, K. Scott, C. Stanton, K.S. Swanson, P.D. Cani, K. Verbeke, G. Reid, Expert consensus document: the international scientific Association for Probiotics and Prebiotics (ISAPP) consensus statement on the definition and scope of prebiotics, Nat. Rev. Gastroenterol. Hepatol. 14 (8) (2017) 491–502, https://doi.org/10.1038/NRGASTRO.2017.75.

[36] Y. Baba, M. Iwatsuki, N. Yoshida, M. Watanabe, H. Baba, Review of the gut microbiome and esophageal cancer: pathogenesis and potential clinical implications, Ann. Gastroenterol. Surg. 1 (2) (2017) 99, https://doi.org/10.1002/AGS3.12014.

[37] P. Louis, H.J. Flint, C. Michel, How to manipulate the microbiota: prebiotics, Adv. Exp. Med. Biol. 902 (2016) 119–142, https://doi.org/10.1007/978-3-319-31248-4_9.

[38] R. Vemuri, R. Gundamaraju, R. Eri, Role of lactic acid probiotic bacteria in IBD, Curr. Pharm. Des. 23 (16) (2017), https://doi.org/10.2174/1381612823666170207100025.

[39] F.L.Y. Fong, K.Y. Lam, C.S. Lau, K.H. Ho, Y.H. Kan, M.Y. Poon, H. El-Nezami, E.T.P. Sze, Reduction in biogenic amines in douchi fermented by probiotic bacteria, PLoS One 15 (3) (2020), https://doi.org/10.1371/JOURNAL.PONE.0230916.

[40] G. Bruno, P. Zaccari, G. Rocco, G. Scalese, C. Panetta, B. Porowska, S. Pontone, C. Severi, Proton pump inhibitors and dysbiosis: current knowledge and aspects to be clarified, World J. Gastroenterol. 25 (22) (2019) 2706, https://doi.org/10.3748/WJG.V25.I22.2706.

[41] M.A. Jackson, J.K. Goodrich, M.E. Maxan, D.E. Freedberg, J.A. Abrams, A.C. Poole, J.L. Sutter, D. Welter, R.E. Ley, J.T. Bell, T.D. Spector, C.J. Steves, Proton pump inhibitors alter the composition of the gut microbiota, Gut 65 (5) (2016) 749–756, https://doi.org/10.1136/GUTJNL-2015-310861.

[42] I. Amir, F.M. Konikoff, M. Oppenheim, U. Gophna, E.E. Half, Gastric microbiota is altered in oesophagitis and Barrett's oesophagus and further modified by proton pump inhibitors, Environ. Microbiol. 16 (9) (2014) 2905–2914, https://doi.org/10.1111/1462-2920.12285.

[43] M.E. Cornelius, T.W. Wang, A. Jamal, C.G. Loretan, L.J. Neff, Tobacco product use among adults - United States, 2019, MMWR Morb. Mortal. Wkly Rep. 69 (46) (2020) 1736–1742, https://doi.org/10.15585/MMWR.MM6946A4.
[44] M. Antinozzi, M. Giffi, N. Sini, F. Gallè, F. Valeriani, C. de Vito, G. Liguori, V.R. Spica, M.S. Cattaruzza, Cigarette smoking and human gut microbiota in healthy adults: a systematic review, Biomedicine 10 (2) (2022), https://doi.org/10.3390/BIOMEDICINES10020510/S1.
[45] F. Bishehsari, E. Magno, G. Swanson, V. Desai, R.M. Voigt, C.B. Forsyth, A. Keshavarzian, Alcohol and gut-derived inflammation, Alcohol Res. 38 (2) (2017) 163 (accessed 12 April 2022).
[46] K. Pohl, P. Moodley, A.D. Dhanda, Alcohol's impact on the gut and liver, Nutrients 13 (9) (2021), https://doi.org/10.3390/NU13093170.
[47] E. Vogtmann, R. Flores, G. Yu, N.D. Freedman, J. Shi, M.H. Gail, B.A. Dye, G.Q. Wang, V. Klepac-Ceraj, B.J. Paster, W.Q. Wei, H.Q. Guo, S.M. Dawsey, Y.L. Qiao, C.C. Abnet, Association between tobacco use and the upper gastrointestinal microbiome among Chinese men, Cancer Causes Control 26 (4) (2015) 581, https://doi.org/10.1007/S10552-015-0535-2.
[48] D.S. Michaud, J. Izard, Z. Rubin, I. Johansson, E. Weiderpass, A. Tjønneland, A. Olsen, K. Overvad, M.C. Boutron-Ruault, F. Clavel-Chapelon, L. Dossus, R. Kaaks, V.A. Katzke, H. Boeing, J. Foerster, A. Trichopoulou, A. Naska, G. Ziara, P. Vineis, S. Grioni, D. Palli, R. Tumino, A. Mattiello, P.H.M. Peeters, P.D. Siersema, A. Barricarte, J.M. Huerta, E. Molina-Montes, M. Dorronsoro, J.R. Quirós, E.J. Duell, B. Ohlsson, B. Jeppsson, A. Johansson, P. Lif, K.T. Khaw, N. Wareham, R.C. Travis, T.J. Key, H. Freisling, T. Duarte-Salles, M. Stepien, E. Riboli, H.B. Bueno-De-Mesquita, Lifestyle, dietary factors, and antibody levels to oral bacteria in cancer-free participants of a European cohort study, Cancer Causes Control 24 (11) (2013) 1901–1909, https://doi.org/10.1007/S10552-013-0265-2.
[49] Z. Li, Y. Liu, L. Dou, Y. Zhang, S. He, D. Zhao, W. Zhang, G. Wang, The effects of smoking and drinking on the oral and esophageal microbiota of healthy people, Ann. Transl. Med. 9 (15) (2021) 1244, https://doi.org/10.21037/ATM-21-3264.

CHAPTER 13

Esophageal dysbiosis—Correcting the paradox: Prebiotics, probiotics, or antibiotics?

Kevin V. Houston[a], Alsiddig Elmahdi[b], Isaac Davis[b], Ana Vilela[b], Byung Soo Yoo[b], Steve M. D'Souza[b], and David A. Johnson[c]
[a]Department of Internal Medicine, Virginia Commonwealth University, Richmond, VA, United States
[b]Department of Internal Medicine, Eastern Virginia Medical School, Norfolk, VA, United States
[c]Division of Gastroenterology, Department of Internal Medicine, Eastern Virginia Medical School, Norfolk, VA, United States

Introduction

There has been an expanding broad interest in targeting the microbiome for many gastrointestinal (GI) diseases. This includes irritable bowel syndrome, small intestinal bowel overgrowth (SIBO), inflammatory bowel disease, celiac disease, liver disease, and more recently gastroesophageal reflux disease (GERD). The preponderance of the literature focusing on the influence of the microbiome in esophageal disease has reported disease concordance with reductions in the gram-positive to gram-negative ratio [1]. As the pathogenic role of esophageal microbiota dysbiosis becomes clearer, it is logical to evaluate targeted therapeutic interventions. Potential methods or adverse effects of therapeutic alteration of microbiome may exist, including the use of prebiotics, probiotics, antibiotics, and bacteriocin-based therapies (Fig. 1). We will summarize the currently available research which may support the use of these therapies for select esophageal disease states. Specifically, the focus will be on how therapies may enhance or restore the gram-positive to gram-negative ratio, either directly or indirectly. Many studies have demonstrated that these disease states may be initiated, and further driven, when gram-negative organisms become overly abundant.

Prebiotics

One such intervention that has been studied and shows promising results comes in the form of prebiotics. In this section, we will briefly review the definition and properties of prebiotics (Table 1).

Esophageal Disease and the Role of the Microbiome
https://doi.org/10.1016/B978-0-323-95070-1.00005-9

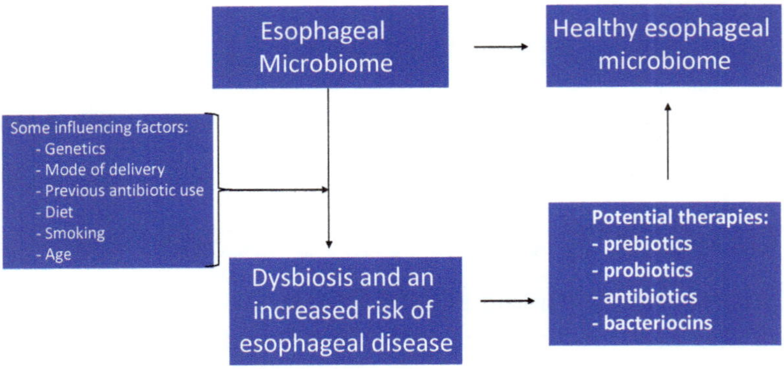

Fig. 1 The esophageal microbiome. Key point: The esophageal microbiome changes are influenced by external factors and changes can be associated with various esophageal diseases. Several potential therapies have been proposed to help transition the microbiome into a healthier state.

Table 1 Therapies comparison.

Prebiotics [2]	Probiotics [3]	Bacteriocins [4]
Substrate that is selectively used by host microorganisms conferring a health benefit	Live organisms, which when administered in adequate amounts, confer a health benefit on the host	Ribosomal-derived peptides
Examples: Human milk oligosaccharides (HMOs)	Examples: *Lactobacillus* species	Example: Nisin
Shown to enhance epithelial cell barrier function, decrease pathologic bacterial adherence and biofilm formation, and possess antiinflammatory properties.	Shown to have antiinflammatory properties, potentially inactive carcinogens and reactive oxygen species.	Shown to have antimicrobial properties, antibiofilm properties, and can work synergistically with conventional drugs, and has immunomodulatory role with cancer therapies.

Key point: Potential methods or adverse effects of therapeutic alteration of microbiome may exist, including the use of prebiotics, probiotics, and bacteriocin-based therapies.

In 2017, the International Scientific Association for Probiotics and Prebiotics (ISAPP) published an updated consensus statement regarding the definition and scope of prebiotics. Per this statement, a prebiotic is "a substrate that is selectively utilized by host microorganisms conferring a health benefit" [2]. As compared with other gut altering substances such as antibiotics and probiotics, prebiotics can be distinguished by selective utilization. In other words, a prebiotic is a nonviable substance with nutritional benefit biased for certain host microorganisms that, when metabolized, imparts a well-defined health benefit or symptom improvement for the host. These benefits may be the result of the prebiotic directly increasing the concentration of specific microorganisms or conferred through utilization of byproducts by microorganisms other than the ones that originally metabolized the substrate, a process referred to as bacterial cross-feeding [2].

Currently, based on the definition set forth by the ISAPP, the only substances extensively studied that meet the proposed criteria are carbohydrate-based fibers. Accepted prebiotics fibers include human milk oligosaccharides (HMOs), the nondigestible oligosaccharides fructans and galactans, as well as lactulose. Other fibers that have high potential for prebiotic effect are referred to as candidate prebiotic fibers and include polysaccharides such as arabinoxylans (AXOS) and xylo-oligosaccharides. Finally, dietary fibers that have been determined to have prebiotic properties are maltodextrins, raffinose, hemicellulose, cellulose, and pectins, among others [2].

There have been numerous studies that have demonstrated beneficial health outcomes related to the use of prebiotics and their ability to restore or enhance the normal GI microbiota, especially distal to the esophagus [5,6]. Unfortunately, except for a few small trials in GERD patients, there have been, to the best of the authors' knowledge, no human studies conducted to evaluate the effects of prebiotics as they relate specifically to the EM [7,8].

Gastroesophageal reflux disease (GERD), Barrett's esophagus (BE), and esophageal adenocarcinoma (EAC)

Early investigations into the EM and dysbiosis demonstrated that all stages in the GERD-BE-EAC pathway share commonalities with regards to the composition in the gastroesophageal (GE) flora. Broadly speaking, the normal GE microbiome (type I) is characterized by a predominance of gram-positive organisms, specifically *Streptococcus* spp., whereas the diseased esophagus (type II) is characterized by a relatively higher proportion of gram-negative anaerobes/microaerophiles [9]. The dysbiosis that occurs due to changing from a type I to type II microbiome is now well documented to be associated with the initial inflammatory process that occurs in GERD and BE and persists in EAC, namely the LPS-TLR4-NF-κB pathway [10]. Further research has suggested that the specific predominant gram-negative bacterial species differs in each step along this pathway and may play an important role in pathogenesis [11].

A case series study published in 2018 demonstrated that daily oral ingestion of a prebiotic maltosyl-isomaltooligosaccharides (MIMO) reduced or eliminated GERD symptoms in 21 of 24 participants after 2–8 weeks of continued use [7]. Previous research identified that isomaltooligosaccharides have been shown to selectively target and increase the concentration of the gram-positive bacteria *Lactobacillus* and *Bifidobacterium* spp. within the human colon [12]. In addition to the colon, *Lactobacillus* are known colonizers of the human oral cavity as well as the esophagus. Therefore, the authors theorized that symptom reduction in GERD patients using MIMO may be related to an increase in oral and esophageal concentration of *Lactobacillus*. An alternative hypothesis is that MIMO in addition to *Lactobacillus* may also increase the abundance of *Streptococcus* spp. The idea for this comes from a study by Deshpande et al. that characterized the phenotypically normal esophagus as "esotype 2" and dominated by an abundance of *Streptococcus* spp. They also noted that each esotype has a distant functional signature and noted esotype 2 as being enriched for the pentose phosphate pathway as well as fructose and mannose metabolism [13]. A tolerability study for MIMO in patients with GERD was completed in August of 2021 (NCT04491734), and a prospective randomized trial was reported to start in 2022, but no results are currently available.

A double-blind, randomized, placebo-controlled study by Beckett demonstrated promising results for the use of prebiotic whole plant sugarcane flour (PSCF) in reducing the symptoms of GERD [8]. A total of 43 participants were randomized in this pilot study that compared daily consumption of 3 g PSCF ($n=22$) and to a placebo cellulose ($n=21$) over the course of 3 weeks. The severity of GERD symptoms was assessed using the validated Gastroesophageal Reflux Disease-Health Related Quality of Life (GERD-HRQL) questionnaire before and after the treatment period. This questionnaire evaluates individual heartburn and regurgitation symptoms and provides a composite score. After 3 weeks of intervention, there were significant differences between the PSCF and placebo cellulose group in the classifications of symptoms (improved, unchanged, or worsened) and overall questionnaire scores. A total of 65% of those in the PSCF group reported improvement in their total score with 55% and 65% indicating improvement in heartburn and regurgitation symptoms, respectively. However, although most subjects in the PSCF group noted improvement of symptoms, there were also 25%, 35%, and 30% of who reported worsening heartburn, regurgitation, and total scores, respectively. Although the authors discussed several potential reasons for why PSCF could improve GERD symptoms, including gut mucosal adherence, gastric motility, and reducing bioavailable nitric oxide, they did not discuss the implications for PSCF as it relates to the GI microbiome, esophageal or otherwise [8]. However, the statistically significant, albeit mixed results among the treatment group, could be confounded by confounding factors in a small sample size and inter-individual variation of the metabolic processing effects. For example, although there was not a statistically significant difference in this study, difference in age of participants (51.1 ± 10.4) and that 40% were smokers could be playing a large role as these two characteristics have been shown to be key determinants of the EM composition, as first reported by Deshpande [13].

One of the first substances recognized for their prebiotic effects are the oligosaccharides present in human milk. HMOs have been shown to profoundly shape the development of a newborn's intestinal microbiota and immune system which carry lifelong implications [14]. When consumed by infants, HMOs increase the concentration of gram-positive Bifidobacteriaceae and Bacteroidaceae present in the GI tract [2]. Unfortunately, the *Bifidobacterium* spp. (*B. longum* subsp. longum, *B. adolescentis*, and *B. lactis*) within the adult GI tract are not equipped with the full repertoire of enzymes necessary to degrade HMOs into functional subunits that can be used by the adult GI microbiome [2].

Researchers have developed synthetic compounds derived from HMOs, some of which have promising prebiotic effects for the adult. The first HMO supplementation tested in adults using a parallel, double-blind, randomized, placebo-controlled study demonstrated clear prebiotic effects of 2′-*O*-fucosyllactose (2′FL) and lacto-*N*-neotetraose (LNnT) after 2 weeks of daily consumption. Microbiota profiling of participant fecal samples showed significant increases in the relative abundance of *Bifidobacteria* (to >25% increase in some participants) and a significant reduction in the Firmicutes and Proteobacteria phyla. Responses were dose dependent, although independent of initial *Bifidobacteria* concentrations. This is an important comparison to galactooligosaccharides, compounds which demonstrate the greatest bifidogenic response in those who have an initially high abundance [2]. Given that each phase of the GERD-BE-EAC pathway has been shown to correlate with a relative increase of Proteobacteria [1], this study provides implications for the use of HMOs and their derivatives, as a prebiotic for EM-based therapy.

The potential use for HMOs in esophageal disease may extend beyond their ability to act as a prebiotic. Some human studies have demonstrated that HMOs can enhance epithelial cell barrier function, decrease pathogenic bacterial adherence and biofilm formation, as well as augment cell absorptive properties [15]. In addition, HMOs have even been shown to possess anti-inflammatory properties. Sodhi et al. demonstrated protective effects of 2′-fucosyllactose (2′-FL) and 6′-sialyllactose (6′-SL) against the development of necrotizing enterocolitis in both murine humans. Cardinal to this protection is the demonstrated effect of both 2′-FL and 6′-SL can directly bind to the lipopolysaccharide (LPS) docking pocket of TLR4, thereby preventing its activation and subsequent inflammatory signaling cascade [16]. The LPS is a "shell" around gram-negative organisms that hardens them to variation in pH, bile salt concentration, proteases, and multiple endemic host immunologic protective pathways [17]. This induces the gram-negative inflammatory cascades and the related downstream adverse consequences [18,19]. Taken together, the studies on HMOs show significant promise and may pack a one–two punch in battling esophageal disease related to microbial dysbiosis and LPS-TLR4-related inflammation.

Eosinophilic esophagitis (EoE)

The specific role that the microbiome plays in the pathophysiology of EoE has yet to be determined. There is emerging evidence, however, to support a state of dysbiosis, predominated by gram-negative organisms [20]. These include *Proteobacteria* ssp., *Neisseria* spp., and *Haemophilus* spp., with the latter two shown to be particularly abundant in children with active EoE and adults, respectively [20,21].

A recent study using murine models demonstrated that fecal matter transplants (FMT) can restore the normal esophageal microbiota [22]. The normative murine EM demonstrates some similarities to the human esophagus; with both containing large concentrations of Firmicutes (*Lactobacillales*) and Bacteroidetes (*Prevotellaceae* taxa in humans) [22]. The experimental design involved induction of a state of microbial dysbiosis through administration of systemic antibiotics that resulted in depletion of Firmicutes, replicating the dysbiosis seen in patients with EoE [22]. The FMT was then administered orally to the treatment group. Subsequently, the esophageal biopsies of FMT-treated mice were compared with a control group. Samples taken from the treatment group demonstrated restoration of the esophageal microbiota enriched in Firmicutes and specifically *Lactobacillales* spp. [22].

Whether or not this was the result of direct colonization or a prebiotic-like effect of microbial-related metabolic products was not mentioned. In addition, and perhaps more importantly, researchers found that the antibiotic administration profoundly diminished the relative abundance of Firmicutes *Lactobacillales* spp. and created a significant increase in the population of *Proteobacteria* and *Actinobacteria*, specifically *Mycoplasmataceae* spp. This shift was found to be associated with pathologic changes in tissue morphology and gene expression, with exaggeration of Type 2 inflammation [22]. The demonstration that FMT, with predefined concentrations of beneficial gram-positive flora, would suggest further research with human studies in attempts to recolonize the esophagus with commensal flora is warranted.

Limitations to the use and study of prebiotics

One of the limitations to identifying prebiotics for potential therapeutic modulation of the esophageal microbiomes is that certain substances have been shown to have prebiotic properties in some populations, but not others [23]. Notably, prebiotics must target a predefined type of organisms, albeit subject to potential alterations by other influences, including environmental, geographic location, and diet, all of which may potentially alter the microbiome. Another limitation is consolidating the available research in terms of the way in which they report their organisms and whether or not they define the flora beyond phyla level. Nonetheless, the evidence supports prebiotics as a promising therapeutic option to favorably effect antiinflammatory or immune modulators.

Probiotics

Probiotics (Table 1) are defined by the Food and Drug Administration (FDA) as "live organisms, which when administered in adequate amounts, confer a health benefit on the host" [3]. Mechanistically, probiotics influence GI disease through a variety of proposed mechanisms. This includes shifting the makeup of the microbiome, competing with harmful bacteria and enhancing mucosal integrity [3]. Other proposed mechanisms include regulating the innate and adaptive immune responses and reducing luminal pH in the colon [3].

Lactobacillus species have been the mainstay of clinical trials showing a capacity to suppress several diseases in the GI tract, ranging from GERD symptom to carcinoma of the esophagus, stomach, and colon [1,24]. Probiotic administration of *Lactobacillus rhamnosus*, *Escherichia coli* Nissle 1917 and heat-inactivated VSL#3 (probiotic food) has been shown to inhibit angiogenesis and inflammation in hepatocellular carcinoma (HCC) with a shift toward *Prevotella* ssp. and *Oscillibacter* ssp., which produce antiinflammatory products [25]. In addition, probiotics have also shown a capacity to preserve biome integrity by shifting toward drug-resistant bacteria in triple therapy when treating *Helicobacter pylori* [26].

Despite the relative novelty of understanding of the esophageal microbiome, the role of probiotics has been investigated here as well. Through animal studies and genomic analysis, probiotics have demonstrated capacity to favorably influence the esophageal microbiome and inflammatory cascade as well as potentially to inactivate carcinogens and reactive oxygen species and will be discussed below.

Gastroesophageal reflux disease

GERD is a common inflammatory disease classically caused by gastric acid reflux. The two major phenotypes of GERD are reflux esophagitis (RE) and nonerosive reflux disease (NERD).

Despite the original paradigm of pathogenesis, recent evidence suggests that GERD esophagitis is mediated primarily by cytokine release [27]. Acid exposure is believed to stimulate epithelial cells resulting in increased expression of IL-8 and other chemokines to promote leukocyte migration [18]. In addition, acid-mediated epithelial damage induces expression of calcitonin gene-related peptide and substance P leading to further cytokine release, local inflammation, and hydrogen peroxide production, which reduces tone of the lower esophageal sphincter [28,29].

The phenotype of RE is associated with decreased Firmicutes and increases in the gram-negative *Fusobacteria* and *Proteobacteria*, whereas NERD is associated with increased proportions of *Proteobacteria*, *Bacteroidetes*, and *Prevotella* and decreased *Fusobacteria* and *Actinobacteria* [11]. This shift in the microbial community may play a role in the inflammatory response seen in GERD. LPS and other Gram-negative released products bind to TLR-4 stimulating an interleukin cascade that ultimately results in TNF production and esophageal sphincter relaxation which increases acid reflux and perpetuates the inflammatory cycle [1].

In addition to their use in regulation of the intestinal biome, probiotics have shown utility in treating the symptoms of GERD [30]. In 2018, a prospective double-blinded randomized placebo-controlled trial of 100 healthy Japanese adults showed significantly higher relief of postprandial discomfort when patients consumed 100 mL of *Bifidobacterium bifidum* 10,347 fermented milk [31]. More recently, a systematic review of 13 prospective studies showed similar results [32], and of the 14 comparison studies, 11 reported positive benefits of probiotics. One of the utilized studies in this systematic review, 79% of participants reported improved GERD symptoms with probiotics. In another study, 81% of participants reported improved nausea, abdominal pain, belching or burping and 45% reported decreased reflux [32].

Another role for probiotics involves their capacity for modulating the deleterious effects of proton pump inhibitors on the microbiome. In 2019, a randomized control trial of 134 RE showed that taking a combination of *Bacillus subtilis* and *Enterococcus faecium* was associated with resolution of SIBO in patients taking 20 mg esomeprazole BID. Moreover, the probiotics were associated with a significant reduction in relapse and time to relapse ($P = .041$), suggested through their discussion to be attributed to a restoration of the microbiome diversity [30].

Barrett's esophagus

Barrett's esophagus (BE) is characterized by metaplastic transformation from stratified squamous to columnar epithelium (intestinal metaplasia) in the setting of longstanding GERD. BE is associated with a Type 2 microbiome consisting of increased *Fusobacteria* and *Proteobacteria* and decreased overall microbial diversity as well as decreases in *Bacteroides* and *Prevotella* [9]. This microbial profile is believed to play a role in the pathology of BE through gram negative–mediated LPS-TLR4-NFkB inflammatory pathway [18].

Studies have demonstrated the potential for probiotics to modulate the expression of BE biomarkers. In an in vitro study, BE cells co-cultured with *Bifidobacterium longum* (*B. longum*) and *Lactobacillus acidophilus* demonstrated reduced expression of IL-18, TNFa, COX2, and CDX1. These results were observed when probiotics were given therapeutically as well as prophylactically; however, the effect was greater with therapeutic administration [33]. These data support the notion that Probiotics can reduce inflammation as well as suppress mutagenesis in the treatment of BE. Furthermore, probiotic administration may play a role in preventing progression of the GERD-BE-Adenocarcinoma sequence. Further evidence for this will be discussed in the next section.

Cancer

The relationship between the esophageal microbiome and cancer is a complex one, particularly in evaluating efficacy of probiotic use. Not only is carcinogenesis itself associated with inflammatory changes observed in those with microbiome dysbiosis, but chemotherapeutic agents are associated with increased susceptibility to infection and microbiome dysregulation. Furthermore, microbiome dysbiosis has been attributed to carcinogenesis and immune dysregulation. Because studies have demonstrated a capacity for probiotics to suppress colorectal carcinoma through mechanisms influencing the innate immune response, oxidative stress, and the composition of the gut microbiome, there is merit in exploring their influence on upper GI cancers [24]. For example, in the stomach, fermented lactobacillus and streptococcus have been shown to degrade pro-carcinogens [34]. The rest of this section will review the data available on probiotic use on esophageal carcinoma and its treatment.

EAC is a possible sequela of BE that is associated with increased *Proteobacteria* and decreased *Firmicutes* as well as the activation of the LPS-TLR4-NF-kB cascade described in GERD and BE. Furthermore, accentuation of this cascade results in increased cyclooxygenase 2 (COX-2), which is associated with progression from BE to high-grade dysplasia and EAC [1]. In addition to esophageal dysbiosis, changes in the gastric microbiome have also been shown to increase the incidence of EAC [35]. To date, no studies have been conducted on probiotic treatment in EAC specifically. However, the previously mentioned impact of probiotic treatment on the production of COX-2 provides support for a probiotic role in prevention.

Although EAC is the most common esophageal cancer in the western world, esophageal squamous cell carcinoma (ESCC) is a larger problem in nondeveloped countries [36]. It is associated with several risk factors, including consumption of alcohol and processed foods as well as changes in the microbiome [36]. Of note, the negative effects of alcohol consumption on the colonic microbiome such as increased luminal pH, decreased *Fusobacterium*, and increased *Proteobacteria* have been shown to be reversible with *Lactobacilli* administration [37]. Because ESCC is associated with decreased *Fusobacterium* and alcohol consumption as a risk factor, it is reasonable to explore whether there is a similar function of probiotics in the esophagus for prevention of ESCC [38].

The risk of ESCC from consumption of processed foods involves the carcinogenesis of biogenic amines such as those associated with Chinese fermented foods [34]. Combination of *Lactobacilli* species, *Lactobacillus rhamnosus* GG (LGG), and *Lactobacillus casei Shirota* (LcS) as well as *E. coli* Nissle 1917 (EcN) to Chinese fermented food has been shown to significantly reduce production of tyramine, a carcinogenic bioamine [34]. Further studies are needed to elicit efficacy of these probiotics use for patients at high risk for ESCC.

Beyond cancer prevention, some studies have shown that probiotics may also play a role in supporting patients undergoing treatment for esophageal carcinoma. For example, a randomized study showed decreased chemotherapy side effects of diarrhea and lymphopenia ($P = .033$, .035, respectively) in esophageal cancer patients taking neoadjuvant synbiotics [39]. Other studies have shown that perioperative symbiotic, combined pro- and prebiotics, decreased incidence of postop infection and abdominal symptoms as well as bacterial translocation in mesenteric lymph nodes in esophageal cancer patients undergoing esophagectomy [40,41]. These events are likely due to competitive influence of probiotics on translocating bacteria and dysbiosis that occur post procedure.

Candidiasis

Candida albicans is an opportunistic fungus that causes mucosal candidiasis, a disease closely associated with GI inflammation. Oral and esophageal candidiasis are most commonly seen in immunocompromised patients. Oral administration of combined *Lactobacilli* species (*L. helveticus* and *L. gasseri*) can activate macrophages toward a fungicidal phenotype, reducing inflammation and preserving murine gastric and esophageal biome integrity [42]. Replication in human tissue specimens is warranted in the investigation of probiotic utility.

In addition to bacterial organisms, yeast probiotics have also shown a capacity for effectiveness against oropharyngeal candidiasis with early resolution of the inflammatory response [43]. Therefore, the use of yeast such as *Saccharomyces cerevisiae* is worth exploring as a potential therapeutic agent in esophageal candidiasis.

Eosinophilic esophagitis

EoE is a common non-IgE-mediated atopic disorder of the esophagus. It is characterized by chronic, immune-mediated esophageal dysfunction and Th2 cell-mediated eosinophilic inflammation of the esophageal mucosa [44]. Along with genetic factors, EoE can be triggered by allergens as well as disruption of the microbiome. The proposed inflammatory cascade involves the markers IL-5 and IL-13. It is postulated that gram-positive *Clostridia* species prevent sensitization to specific food allergens inducing inflammation in patients with genetic predispositions. Along with decreased *Clostridia*, EoE is associated with decreased *Bacteroides* and increased *Proteobacteria* and *Corynebacterium* [44].

The benefits of probiotics in EoE have been demonstrated in murine models. *Lactococcus lactis* NCC 2287 significantly reduces esophageal eosinophilia when given therapeutically [45]. However, no response was observed when the probiotic was given prophylactically [45]. Replication of this therapy can make a strong case for the use of probiotics in the desensitization of allergens in EoE, especially because similar results have been demonstrated when administering *Lactococcus* ssp. in other atopic diseases [46].

Antibiotics

Intraluminal antibiotics are a well-established treatment modality in many GI diseases, including small intestinal bacterial overgrowth (SIBO), hepatic encephalopathy, and *Clostridium difficile* infection. Because of the potential of these agents to promote dysbiosis, and resistance of gram-negative bacteria, there is a potential for use to increase the risk of esophageal disease [17,47,48]. There is currently no substantial data on the association of antibiotics with GERD, BE, or other "noninfectious" diseases of the esophagus. There is conflicting animal model evidence on whether or not chronic antibiotic use increases the risk of esophageal carcinoma [49,50]. Interestingly, one study looking at a group of patients who received antibiotics within 60 days of chemotherapy initiation vs those who did not receive antibiotics showed that the rate of primary progressive disease in the antibiotic-treated group was significantly higher [51]. Although unclear from their preliminary data, they attributed this to the antibiotic microbiota dysbiosis and suggested more studies are warranted to confirm the effect as well as identify the specific microbiota species involved [17].

Bacteriocins

Bacteriocins (Table 1) are ribosomal-derived peptides that have been divided in two classes, the first are modified enzymatically (class I bacteriocins) and the second that remain unaltered. They are positively charged and are smaller than 10 kDa [4,52]. Their primary application has been in the food preservation industry for past decades, as time has passed, new variants have been studied, especially for their antimicrobial properties. For example, a well-studied bacteriocin called Nisin, produced by gram-positive bacteria, has shown to have antimicrobial activity against gram-positive and -negative bacteria, in addition to antibiofilm properties, and can work synergistically with conventional drugs [53,54].

In addition, bacteriocins have been studied to have an immunomodulatory role, with potential use in cancer therapy. Nisin, a bacteriocin produced by the dairy starter culture bacterium, *Lactococcus lactis* ssp., demonstrated to have potential antitumor effects in head and neck squamous cell carcinoma cells [55]. It was demonstrated in vivo that this is achieved through different mechanisms, such as via the induction of apoptosis, in squamous tumor and endothelial cells, as well as inhibition of tumor cell proliferation [55].

Literature describes that bacteriocins are produced in the GI tract and protect the host against infection, and a study postulates that bacteriocins are likely to interact with the epithelial cells and cross the gut blood barrier [56]. At present, it is unknown to what extent by crossing this barrier bacteriocins can achieve a beneficial effect (such as the antibacterial or immunomodulatory) in the esophagus or elsewhere in the GI tract [52,57].

Currently, research into the bacteriocin's role in the esophagus microbiome is limited, but certainly the ongoing research on its antimicrobial properties and its antineoplastic role opens a wide array of future applications in this area.

Conclusions

In conclusion, studies have begun to support the esophageal microbiome playing a major role in the pathogenesis of esophageal disease [1,58]. Various factors can affect the composition of the commensal flora, which may lead to dysbiosis and ultimately result in esophageal diseases. Potential disease mitigation strategies or adverse effects of therapeutic alteration of microbiome may exist, including the use of prebiotics, probiotics, antibiotics, and bacteriocin-based therapies. More data are needed, although there is emerging evidence that these concepts may open new pathways for redefining pathogenesis and directed treatment in a wide range of esophageal diseases.

Future directions

- In specific populations, prebiotics may play a role in stimulating the healthy bacteria in the gut and may play a role in a healthy esophageal microbiome. Defining specifics of which prebiotics, probiotics, antibiotics, and bacteriocins have beneficial effect will be critical.

- Better understanding of any adverse effects of these therapies will be important.
- Better definition of the mechanistic pathways by which any of these therapies will be used is key.
- Randomized controlled trials with well-designed methodology to evaluate microbial and cytokine effects are needed.
- Evaluation of these therapies as a single therapy versus adjunctive to current traditional therapies will need to be assessed.

Author contributions

Dr. DA Johnson contributed to the construction of the project; all authors wrote and edited the manuscript.

Potential conflict of interest

DA Johnson is a consultant and clinical investigator for ISOthrive.

References

[1] S.M. D'Souza, K. Houston, L. Keenan, B.S. Yoo, P.J. Parekh, D.A. Johnson, Role of microbial dysbiosis in the pathogenesis of esophageal mucosal disease: a paradigm shift from acid to bacteria? World J. Gastroenterol. 27 (18) (2021) 2054–2072, https://doi.org/10.3748/wjg.v27.i18.2054.
[2] G.R. Gibson, R. Hutkins, M.E. Sanders, et al., Expert consensus document: the International Scientific Association for Probiotics and Prebiotics (ISAPP) consensus statement on the definition and scope of prebiotics, Nat. Rev. Gastroenterol. Hepatol. 14 (8) (2017) 491–502, https://doi.org/10.1038/nrgastro.2017.75.
[3] J. Lambert, R. Hull, Upper gastrointestinal tract disease and probiotics, Asia Pac. J. Clin. Nutr. 5 (1) (1996) 31–35.
[4] S. Heilbronner, B. Krismer, H. Brötz-Oesterhelt, A. Peschel, The microbiome-shaping roles of bacteriocins, Nat. Rev. Microbiol. 19 (11) (2021) 726–739, https://doi.org/10.1038/s41579-021-00569-w.
[5] A. Gagliardi, V. Totino, F. Cacciotti, et al., Rebuilding the gut microbiota ecosystem, Int. J. Environ. Res. Public Health 15 (8) (2018), https://doi.org/10.3390/ijerph15081679.
[6] M.E. Sanders, D.J. Merenstein, G. Reid, G.R. Gibson, R.A. Rastall, Probiotics and prebiotics in intestinal health and disease: from biology to the clinic, Nat. Rev. Gastroenterol. Hepatol. 16 (10) (2019) 605–616, https://doi.org/10.1038/s41575-019-0173-3.
[7] J. Selling, P. Swann, L.R. Madsen, J. Oswald, Improvement in gastroesophageal reflux symptoms from a food-grade maltosyl-isomaltooligosaccharide soluble fiber supplement: a case series, Integr. Med. (Encinitas) 17 (5) (2018) 40–42.
[8] J.M. Beckett, N.K. Singh, J. Phillips, et al., Anti-heartburn effects of sugar cane flour: a double-blind, randomized, placebo-controlled study, Nutrients 12 (6) (2020), https://doi.org/10.3390/nu12061813.

[9] L. Yang, X. Lu, C.W. Nossa, F. Francois, R.M. Peek, Z. Pei, Inflammation and intestinal metaplasia of the distal esophagus are associated with alterations in the microbiome, Gastroenterology 137 (2) (2009) 588–597, https://doi.org/10.1053/j.gastro.2009.04.046.
[10] J. Lv, L. Guo, J.J. Liu, H.P. Zhao, J. Zhang, J.H. Wang, Alteration of the esophageal microbiota in Barrett's esophagus and esophageal adenocarcinoma, World J. Gastroenterol. 25 (18) (2019) 2149–2161, https://doi.org/10.3748/wjg.v25.i18.2149.
[11] J. Zhou, P. Shrestha, Z. Qiu, et al., Distinct microbiota dysbiosis in patients with non-erosive reflux disease and esophageal adenocarcinoma, J. Clin. Med. 9 (7) (2020), https://doi.org/10.3390/jcm9072162.
[12] C.H. Yen, Y.H. Tseng, Y.W. Kuo, M.C. Lee, H.L. Chen, Long-term supplementation of isomalto-oligosaccharides improved colonic microflora profile, bowel function, and blood cholesterol levels in constipated elderly people—a placebo-controlled, diet-controlled trial, Nutrition 27 (4) (2011) 445–450, https://doi.org/10.1016/j.nut.2010.05.012.
[13] N.P. Deshpande, S.M. Riordan, N. Castaño-Rodríguez, M.R. Wilkins, N.O. Kaakoush, Signatures within the esophageal microbiome are associated with host genetics, age, and disease, Microbiome 6 (1) (2018) 227, https://doi.org/10.1186/s40168-018-0611-4.
[14] D. Garrido, S. Ruiz-Moyano, D.G. Lemay, D.A. Sela, J.B. German, D.A. Mills, Comparative transcriptomics reveals key differences in the response to milk oligosaccharides of infant gut-associated bifidobacteria, Sci. Rep. 5 (2015) 13517, https://doi.org/10.1038/srep13517.
[15] V. Triantis, L. Bode, R.J.J. van Neerven, Immunological effects of human milk oligosaccharides, Front. Pediatr. 6 (2018) 190, https://doi.org/10.3389/fped.2018.00190.
[16] C.P. Sodhi, P. Wipf, Y. Yamaguchi, et al., The human milk oligosaccharides 2′-fucosyllactose and 6′-sialyllactose protect against the development of necrotizing enterocolitis by inhibiting toll-like receptor 4 signaling, Pediatr. Res. 89 (1) (2021) 91–101, https://doi.org/10.1038/s41390-020-0852-3.
[17] Z. Breijyeh, B. Jubeh, R. Karaman, Resistance of gram-negative bacteria to current antibacterial agents and approaches to resolve it, Molecules 25 (6) (2020), https://doi.org/10.3390/molecules25061340.
[18] R.F. Souza, X. Huo, V. Mittal, et al., Gastroesophageal reflux might cause esophagitis through a cytokine-mediated mechanism rather than caustic acid injury, Gastroenterology 137 (5) (2009) 1776–1784, https://doi.org/10.1053/j.gastro.2009.07.055.
[19] P.C. Konturek, A. Nikiforuk, J. Kania, M. Raithel, E.G. Hahn, S. Mühldorfer, Activation of NFkappaB represents the central event in the neoplastic progression associated with Barrett's esophagus: a possible link to the inflammation and overexpression of COX-2, PPARgamma and growth factors, Dig. Dis. Sci. 49 (7–8) (2004) 1075–1083, https://doi.org/10.1023/b:ddas.0000037790.11724.70.
[20] A.J. Benitez, C. Hoffmann, A.B. Muir, et al., Inflammation-associated microbiota in pediatric eosinophilic esophagitis, Microbiome 3 (2015) 23, https://doi.org/10.1186/s40168-015-0085-6.
[21] J.K. Harris, R. Fang, B.D. Wagner, et al., Esophageal microbiome in eosinophilic esophagitis, PLoS One 10 (5) (2015), e0128346, https://doi.org/10.1371/journal.pone.0128346.
[22] M. Brusilovsky, R. Bao, M. Rochman, A.M. Kemter, C.R. Nagler, M.E. Rothenberg, Host-microbiota interactions in the esophagus during homeostasis and allergic inflammation, Gastroenterology 162 (2) (2022) 521–534.e8, https://doi.org/10.1053/j.gastro.2021.10.002.
[23] L. Zhang, X. Gu, J. Wang, et al., Effects of dietary isomaltooligosaccharide levels on the gut microbiota, immune function of sows, and the diarrhea rate of their offspring, Front. Microbiol. 11 (2020), 588986, https://doi.org/10.3389/fmicb.2020.588986.

[24] C. Meng, C. Bai, T.D. Brown, L.E. Hood, Q. Tian, Human gut microbiota and gastrointestinal cancer, Genom. Proteom. Bioinform. 16 (1) (2018) 33–49, https://doi.org/10.1016/j.gpb.2017.06.002.
[25] J. Li, C.Y.J. Sung, N. Lee, et al., Probiotics modulated gut microbiota suppresses hepatocellular carcinoma growth in mice, Proc. Natl. Acad. Sci. U. S. A. 113 (9) (2016) E1306–E1315, https://doi.org/10.1073/pnas.1518189113.
[26] B. Oh, B.S. Kim, J.W. Kim, et al., The effect of probiotics on gut microbiota during the *Helicobacter pylori* eradication: randomized controlled trial, Helicobacter 21 (3) (2016) 165–174, https://doi.org/10.1111/hel.12270.
[27] K.B. Dunbar, A.T. Agoston, R.D. Odze, et al., Association of acute gastroesophageal reflux disease with esophageal histologic changes, JAMA 315 (19) (2016) 2104–2112, https://doi.org/10.1001/jama.2016.5657.
[28] L. Cheng, K.M. Harnett, W. Cao, et al., Hydrogen peroxide reduces lower esophageal sphincter tone in human esophagitis, Gastroenterology 129 (5) (2005) 1675–1685, https://doi.org/10.1053/j.gastro.2005.09.008.
[29] A. Ustaoglu, A. Nguyen, S. Spechler, D. Sifrim, R. Souza, P. Woodland, Mucosal pathogenesis in gastro-esophageal reflux disease, Neurogastroenterol. Motil. 32 (12) (2020), e14022, https://doi.org/10.1111/nmo.14022.
[30] Q.H. Sun, H.Y. Wang, S.D. Sun, X. Zhang, H. Zhang, Beneficial effect of probiotics supplements in reflux esophagitis treated with esomeprazole: a randomized controlled trial, World J. Gastroenterol. 25 (17) (2019) 2110–2121, https://doi.org/10.3748/wjg.v25.i17.2110.
[31] A. Gomi, K. Yamaji, O. Watanabe, et al., *Bifidobacterium bifidum* YIT 10347 fermented milk exerts beneficial effects on gastrointestinal discomfort and symptoms in healthy adults: a double-blind, randomized, placebo-controlled study, J. Dairy Sci. 101 (6) (2018) 4830–4841, https://doi.org/10.3168/jds.2017-13803.
[32] J. Cheng, A.C. Ouwehand, Gastroesophageal reflux disease and probiotics: a systematic review, Nutrients 12 (1) (2020), https://doi.org/10.3390/nu12010132.
[33] B. Mozaffari Namin, N.E. Daryani, A. Mirshafiey, M.K.S. Yazdi, M.M.S. Dallal, Effect of probiotics on the expression of Barrett's oesophagus biomarkers, J. Med. Microbiol. 64 (Pt 4) (2015) 348–354, https://doi.org/10.1099/jmm.0.000039.
[34] F.L.Y. Fong, K.Y. Lam, C. San Lau, et al., Reduction in biogenic amines in douchi fermented by probiotic bacteria, PLoS One 15 (3) (2020), e0230916, https://doi.org/10.1371/journal.pone.0230916.
[35] M.T. Abreu, R.M. Peek, Gastrointestinal malignancy and the microbiome, Gastroenterology 146 (6) (2014) 1534–1546.e3, https://doi.org/10.1053/j.gastro.2014.01.001.
[36] C.C. Abnet, M. Arnold, W.Q. Wei, Epidemiology of esophageal squamous cell carcinoma, Gastroenterology 154 (2) (2018) 360–373, https://doi.org/10.1053/j.gastro.2017.08.023.
[37] L. Bull-Otterson, W. Feng, I. Kirpich, et al., Metagenomic analyses of alcohol induced pathogenic alterations in the intestinal microbiome and the effect of *Lactobacillus rhamnosus* GG treatment, PLoS One 8 (1) (2013), e53028, https://doi.org/10.1371/journal.pone.0053028.
[38] Q. Wang, Y. Rao, X. Guo, et al., Oral microbiome in patients with oesophageal squamous cell carcinoma, Sci. Rep. 9 (1) (2019) 19055, https://doi.org/10.1038/s41598-019-55667-w.
[39] M. Motoori, M. Yano, H. Miyata, et al., Randomized study of the effect of synbiotics during neoadjuvant chemotherapy on adverse events in esophageal cancer patients, Clin. Nutr. 36 (1) (2017) 93–99, https://doi.org/10.1016/j.clnu.2015.11.008.
[40] Y. Yokoyama, E. Nishigaki, T. Abe, et al., Randomized clinical trial of the effect of perioperative synbiotics versus no synbiotics on bacterial translocation after oesophagectomy, Br. J. Surg. 101 (3) (2014) 189–199, https://doi.org/10.1002/bjs.9385.

[41] K. Tanaka, M. Yano, M. Motoori, et al., Impact of perioperative administration of synbiotics in patients with esophageal cancer undergoing esophagectomy: a prospective randomized controlled trial, Surgery 152 (5) (2012) 832–842, https://doi.org/10.1016/j.surg.2012.02.021.

[42] H. Authier, M. Salon, M. Rahabi, et al., Oral administration of *Lactobacillus helveticus* LA401 and *Lactobacillus gasseri* LA806 combination attenuates oesophageal and gastrointestinal candidiasis and consequent gut inflammation in mice, J. Fungi (Basel) 7 (1) (2021), https://doi.org/10.3390/jof7010057.

[43] E. Roselletti, S. Sabbatini, N. Ballet, et al., *Saccharomyces cerevisiae* CNCM I-3856 as a new therapeutic agent against oropharyngeal candidiasis, Front. Microbiol. 10 (2019) 1469, https://doi.org/10.3389/fmicb.2019.01469.

[44] M. Mennini, R. Tambucci, C. Riccardi, et al., Eosinophilic esophagitis and microbiota: state of the art, Front. Immunol. 12 (2021), 595762, https://doi.org/10.3389/fimmu.2021.595762.

[45] S. Holvoet, R. Doucet-Ladevèze, M. Perrot, C. Barretto, S. Nutten, C. Blanchard, Beneficial effect of *Lactococcus lactis* NCC 2287 in a murine model of eosinophilic esophagitis, Allergy 71 (12) (2016) 1753–1761, https://doi.org/10.1111/all.12951.

[46] L.B. de Lacerda, W.M. Rios, A.P. Masson, et al., Oral administration of Hsp65-producing *Lactococcus lactis* attenuates allergic asthma in a murine model, J. Appl. Microbiol. 130 (6) (2021) 2075–2086, https://doi.org/10.1111/jam.14913.

[47] M. May, J.A. Abrams, Emerging ensights into the esophageal microbiome, Curr. Treat. Options Gastroenterol. 16 (1) (2018) 72–85, https://doi.org/10.1007/s11938-018-0171-5.

[48] V. di Pilato, G. Freschi, M.N. Ringressi, L. Pallecchi, G.M. Rossolini, P. Bechi, The esophageal microbiota in health and disease, Ann. N. Y. Acad. Sci. 1381 (1) (2016) 21–33, https://doi.org/10.1111/nyas.13127.

[49] A. Sawada, Y. Fujiwara, Y. Nagami, et al., Alteration of esophageal microbiome by antibiotic treatment does not affect incidence of rat esophageal adenocarcinoma, Dig. Dis. Sci. 61 (11) (2016) 3161–3168, https://doi.org/10.1007/s10620-016-4263-6.

[50] B. Boursi, R. Mamtani, K. Haynes, Y.X. Yang, Recurrent antibiotic exposure may promote cancer formation—another step in understanding the role of the human microbiota? Eur. J. Cancer 51 (17) (2015) 2655–2664, https://doi.org/10.1016/j.ejca.2015.08.015.

[51] C. Wu, R. Lai, J. Li, et al., Antibiotics modulate chemotherapy efficacy in patients with esophageal cancer, Cancer Manag. Res. 12 (2020) 4991–4997, https://doi.org/10.2147/CMAR.S248130.

[52] L.M.T. Dicks, L. Dreyer, C. Smith, A.D. van Staden, A review: the fate of bacteriocins in the human gastro-intestinal tract: do they cross the gut-blood barrier? Front. Microbiol. 9 (2018) 2297, https://doi.org/10.3389/fmicb.2018.02297.

[53] J.M. Shin, J.W. Gwak, P. Kamarajan, J.C. Fenno, A.H. Rickard, Y.L. Kapila, Biomedical applications of nisin, J. Appl. Microbiol. 120 (6) (2016) 1449–1465, https://doi.org/10.1111/jam.13033.

[54] M.L. Chikindas, R. Weeks, D. Drider, V.A. Chistyakov, L.M. Dicks, Functions and emerging applications of bacteriocins, Curr. Opin. Biotechnol. 49 (2018) 23–28, https://doi.org/10.1016/j.copbio.2017.07.011.

[55] P. Kamarajan, T. Hayami, B. Matte, et al., Nisin zp, a bacteriocin and food preservative, inhibits head and neck cancer tumorigenesis and prolongs survival, PLoS One 10 (7) (2015), e0131008, https://doi.org/10.1371/journal.pone.0131008.

[56] F. Drissi, S. Buffet, D. Raoult, V. Merhej, Common occurrence of antibacterial agents in human intestinal microbiota, Front. Microbiol. 6 (2015) 441, https://doi.org/10.3389/fmicb.2015.00441.

[57] I. Spadoni, E. Zagato, A. Bertocchi, et al., A gut-vascular barrier controls the systemic dissemination of bacteria, Science 350 (6262) (2015) 830–834, https://doi.org/10.1126/science.aad0135.
[58] Y.R. Nobel, E.J. Snider, G. Compres, et al., Increasing dietary fiber intake is associated with a distinct esophageal microbiome, Clin. Transl. Gastroenterol. 9 (10) (2018) 199, https://doi.org/10.1038/s41424-018-0067-7.

Index

Note: Page numbers followed by *f* indicate figures and *t* indicate tables.

H

I

K

L

M

Kolibri 360 GmbH
Ferdinand-Jühlke-Straße 7
99095 Erfurt, Deutschland
produktsicherheit@kolibri360.de